henryville special

solomon's hairwing

len wright skittering caddis

sid neff hairwing

delta wing caddis

flat wing caddis

quill wing caddis

hare's ear or vermont caddis

To Jerry: 3/16/78

Observe & Experiment,
    the Key to learning the game —
    Good casts &
        Sharp rises —
                Larry Solomon

illustrated with line drawings by John Lane

photography by Larry Solomon and George Mowbray

the
caddis
and the
angler

LARRY SOLOMON AND ERIC LEISER

FOREWORD BY LEONARD M. WRIGHT, JR.

STACKPOLE BOOKS

THE CADDIS AND THE ANGLER
Copyright © 1977 by
Larry Solomon and Eric Leiser

Published by
STACKPOLE BOOKS
Cameron and Kelker Streets
P.O. Box 1831
Harrisburg, Pa. 17105

Published simultaneously in Don Mills, Ontario, Canada by Thomas Nelson & Sons, Ltd.

Printed in the U.S.A.

Designed by Joan Stoliar

Library of Congress Cataloging in Publication Data

Solomon, Larry, 1938-
  The caddis and the angler.

  Bibliography: p.
  Includes index.
  1. Fly fishing. 2. Caddis-flies. 3. Flies,
Artificial. I. Leiser, Eric, joint author. II. Title.
SH456.S59 1977   799.1'2   76-58533
ISBN 0-8117-0312-6

To Ernie Maltz,
who was fishing a caddis
before we knew what it was;
and who gave us his years,
his knowledge, and most of all,
his friendship

contents

foreword

Sitting here on a mid-November evening when no, or very few, insects are hatching from streams, there is at least the armchair pleasure of knowing that an entire order of insects will be emerging from the following pages. Emerging from neglect and obscurity rather than from the water surface, it's true. But any kind of hatch is better than none.

With the publication of this book, caddis flies, an order of insects with over 1,000 species in North America alone, will spread their wings and flutter back into the limelight. I say "back into" for good reason: neither the caddis nor its importance to fly fishing is a recent discovery. It's simply a matter of re-discovery.

A hundred years ago and even long before that, the caddis was a familiar friend to fly fishers. Anglers may not have called it by name, but the evidence shows they certainly knew its appearance well. The classic centuries-old, winged wet fly was, and still is a far more accurate representation of a caddis than a mayfly. Remember also that these patterns were almost invariably tied without tails—another clue to their caddis ancestry.

Ancient fishermen knew how to present these flies in a caddis-like manner, too. Dapping with a single fly or dancing the dropper over the surface are two deadly, though nearly forgotten, techniques of imitating the behavior of winged or adult caddis. It is surprising that both these time-tested artificials and the realistic methods of fishing them have been in eclipse for nearly a century; yet such was the power and influence of Halford's dead-drift mayfly doctrine. The caddis is coming back in popularity now, though, and I think it is back to stay. This type of insect is not only numerous, available, nutritious and delicious, but it also has a unique virtue. Its behavior, both as it emerges and as it returns to the river for waterplay or egg-laying, excites trout far more than the statelier motions of other types of insects do.

In the following chapters you will find out more about the caddis by far than you've ever been able to learn from any previous fishing book. Here's everything you always wanted to know about family, genus, species, lifestyles, imitations plus where, when and how to fish them.

Undoubtedly, more learned and detailed entomological studies of the caddis will be published in the years to come. But they will be of more importance to pure science than to angling. This volume, as far as the fly fisher is concerned, will remain the book of the caddis. If you want to become a more versatile, consistent and effective trouter with nymphs, wet flies and dries, read and re-read this book carefully. Then go out and practice what it preaches.

LEONARD M. WRIGHT, JR.

acknowledgments

In our day, good angling books are seldom really the product of a single angler's mind, or even two. The individual contributions to *The Caddis and the Angler* were so numerous that it is hard to determine who actually deserves more credit for its existence—ourselves as authors, or some one hundred other anglers who were themselves already adept at fishing caddis imitations and tying its patterns. The fact is, this book could not have been written as a comprehensive guide to fishing the caddis without the help of many dedicated and very generous anglers.

Many are mentioned in the following pages. Many persons' names, unfortunately, do not appear, only because their contributions were part of casual conversations which took place between anglers over the years, randomly and without any thought of organizing them into book form. And yet, those bits of information stored subconsciously have emerged in *The Caddis and the Angler.* Many were nameless: the angler on the stream, who said, "Try this", or "Here, fish my water", or the flytyer at a forum who unknowingly taught us a new twist. The introductions were not necessary; we are all members of the same fraternity and under certain circumstances names are not as important as our sense of angling kinship and our sharing with each other knowledge for the enjoyment of fly fishing.

To all of you we are indebted.

Yet, we would be remiss if we did not give special mention, for very special reasons, to the following people.

Dr. Oliver S. Flint, Jr., Curator of Neuropterids of the National Museum of Natural History, Smithsonian Institution, whose time, efforts, and patience in assisting the authors so that they could be correct and up-to-date in every entomological detail was invaluable. Without his help, this volume could not have been completed.

Walt Dette and his wife, Winnie, who so graciously allowed us the use of their home in which to perform experiments in raising aquatic insects, and shared with us unique ideas in both flytying and fishing—and allowed us to stay over when it got too late to go home.

Robert H. Boyle, who, in the midst of his twenty-four-hours-a-day, seven-days-a-week battle for clean water allowed us the use of his home reference collection from which we staggered overloaded with armsful of texts and data related to the caddis; and who in addition gave freely of his time.

Bill Claiborne for his assistance with some of the photography.

Ralph Graves, for his excellent photograph of a natural caddis on the jacket.

Dave McNeese, for seining and netting caddis flies and larvae in the streams of Oregon, in addition to writing sheaves of paper on his own findings and contributing them to our efforts.

Jerry Applegarth, also of Oregon, whose willingness to help included a tape cassette recording not only the caddis activity on the rivers of that state, but their beauty as well.

Gary Borger, of Wausau, Wisconsin, for contributing information accumulated during an independent research of caddis flies in the Midwest and for the invitation to "come see for ourselves" some of the fine trout waters in his home state.

Gary LaFontaine, field editor for *Fly Fisherman Magazine,* of Deer Lodge, Montana, for allowing us to use information from his book, *The Challenge of the Trout.*

Charlie Brooks, philosopher and friend, for the pages of handwritten notes relating to his experiences with caddis flies and trout in both Idaho and Montana.

Joan Stoliar, for taking an extraspecial interest in the design of the overall book—and, perhaps, for inflating our egos a bit when we needed it.

John Lane, our illustrator, for doing a fine job, but especially for bearing with our constant changes, resulting in more and more revisions.

Nancy and June, for liking everything we did. For them, we could do no wrong. Theirs were always prejudiced opinions, but more than welcome.

Jerry Hoffnagle, our editor, who became as excited and enthused over the caddis as the authors; his moral support at many times went above and beyond.

Eric Peper, former editor of the Field & Stream Book Club, and a good friend, for sharing freely his thoughts and knowledge.

Sam Melner, still another Oregonian, whose suggestion to the authors, "Why don't you two get together and do a book on caddis . . ." initiated this work almost three years ago.

I had fished for the better part of the morning; now it was almost noon. A good hatch of hendricksons had been in progress daily since my arrival on the weekend, and I had taken and released a satisfying number of trout. But today things were different. A few fish were rising, but with one or two exceptions they seemed indifferent to my imitations.

I tried a few more arm exercises and then thought better of it. A retreat was in order. Besides, even if the trout weren't hungry, I was. I found a large tree with a comfortable trunk and snuggled my back against it. A nearby log needed very little positioning for a foot rest. The ham and swiss sandwich went nicely with the hot coffee and the late spring day was warming up to the right degree of coziness. A pleasant drowsiness began to weigh down on my eyes. I fought against it, but—delightfully—lost the battle.

Half an hour later the hard ground informed me I was much too spoiled by soft living. Mother Earth just wouldn't yield in the same manner and the right places as my orthopedic mattress. I woke up one eye at a time, shifting one foot, then the other. When the pins and needles subsided I searched the stream once more for any sign of a change in attitude on the part of the trout.

Instead I saw another angler standing in the exact spot I had earlier vacated, a tall lean man, wielding an exceptionally long bamboo fly rod. His casting was effortless, free. From where I watched the river, the rod, and the angler were all one: a portrait in motion.

Suddenly he was into a fish—a good fish! Expertly, the trout was brought to net and the hook deftly removed. He straightened again, tucked the rod under his arms, and pulled a pipe from a vest pocket. He turned to put his back to the wind to light it, and I nodded a greeting.

He waded out of the stream toward me.

"Just thought I'd try your water while you were catnapping," he said.

"Why not," I answered. "At least you seem to have found the secret. I've been casting over that fish you just released all morning. What did he take anyway?"

"Caddis," the old angler replied and smiled.

"Caddis?"

"Mmm hmm," he nodded.

I knew from the way he answered he had to be right. And yet, it just didn't penetrate. The hendrickson pattern I had been fishing the last three days had

caddis
an informal introduction

proven successful, and there was no obvious reason to consider a caddis imitation, and yet . . .

"Little bit of a surprise, isn't it?" he said. "Especially when we've had nothing but mayflies for the last few days. Kind of fools you, in a way. You just get used to one thing and . . ."

"How did you know?" I interrupted. "I mean . . . well, to tell you the truth I really don't recall seeing any caddis on the surface, or in the air, for that matter."

"You will now," he answered. "Take a look there . . . and there . . ."

And indeed, there they were. He smiled again.

"This morning it was just a little tougher," he continued. "Trout weren't really after the hatched fly. They were after the pupa mostly. Did you happen to see the way they were rising?"

"Well," I said, "the few that were rising did seem a little frisky. Noisy."

"That's right," he said, "very natural, too. They were chasing the caddis on its way up. Those caddis pupae come off the bottom and hatch so fast, the fish have to hurry to catch them. Even when they don't get them, their momentum sometimes carries them through the surface. That's why you'll see that splashy rise. Good indicator, most of the time. You can learn a lot from just watching the rise forms. Probably the two most important observations you can make when you're trying to match a natural are the silhouette of the insect and the rise form of the trout. That will usually mark the difference between a mayfly hatch and a caddis hatch."

I was a little miffed, but even more curious. It seemed so simple—no wonder my hendrickson ties weren't drawing any rises! And I had some caddis patterns in my vest all the time.

"Don't feel too badly," he said, reading my mind. "It happens to most of us one time or another. Learning a stream, identifying an insect, knowing how the trout will behave . . . well, I've spent most of a lifetime at it and I'm still learning. It's part of the game, and maybe it makes the game more fun.

"Take these caddis, for instance. Took me a long time to learn that this bug is even more important to the trout than the mayfly. The fact is, on many streams at least, the caddis represents half the trout's regular diet. And yet, many anglers still don't understand how unique this little bug is. When they do use a caddis imitation they try to fish it like a mayfly. Oh, it'll take fish that way but not the way it does if it's fished like the real thing!"

He threw his head back and laughed a little; he seemed to have had lots of good fun learning about this caddis. He was now sitting beside me on the log contemplating his pipe, which had gone out.

"Well, what is this caddis we're seeing?" I asked.

"The one that's hatching right now is called *Brachycentrus numerosus*, commonly called a shad fly, or grannom. Most caddis don't have common names, though, like mayflies, so sooner or later you have to become familiar with the different families by their Latin names. But there're really only a few important ones. I suppose if I did some more collecting and a little more studying,

I'd get to know more, but that's not as important as knowing the habits and characteristics of these insects. That's the thing. If you know when an insect will hatch, what it looks like and how it behaves, you can call it anything you want as long as you imitate its color and size, and its actions. The second part is just as important as the first, especially for caddis.

"Caddis are different from mayflies—not just in their shape, but in the way they behave as well. Take a look at that one over there. Look at him trying to get off the water. There, he's off! Didn't waste much time, did he? Most of them won't. He'll be around for a much longer time than a mayfly too; sometimes ten or twenty days, anyway. After he—or she—mates, the female will lay her eggs in the water. Those eggs will turn into larvae that look like grubs. Each larva will build a case, its own private shelter, from small twigs and little bits of vegetation until it's time for pupation—they go into hibernation for a few weeks, like caterpillars, and make their change into adult winged insects inside the case. That's one stage mayflies don't have at all. Once the pupal stage is complete it breaks out of its case and rockets to the surface fast, throwing off its old skin as it approaches the top.

"They're airborne almost immediately—with a few exceptions. Egg, larva, pupa and adult—that's the whole life cycle.

"Now then, what I've described to you holds true for that bug you see flying around out there. Some caddis, like *Rhyacophila*—you probably call them green rock worm, or just green caddis—are what is known as free-living caddis. They don't make a case, at least not until the very last moments prior to pupation. That's why it's sometimes good to fish a larva imitation; those free-livers are good food for trout."

He paused and lit his pipe again.

"Well," he said, "sometimes old men talk a lot. I suspect you'd rather be fishing than listening to me."

His eyes held me, smiling. No way. There were many days left for fishing, seasons, years. This old angler was something special. I was going to get a lesson.

"What I'd like to know," I said, "is the difference in fishing a caddis as opposed to a mayfly. It's not just a matter of changing patterns, I take it."

"No, it's not," he answered, "though that's the first step. It also depends if you're fishing wet or dry. For instance, a larva is fished much in the manner you would a free-floating nymph, down deep and bouncing off the bottom now and again. A pupa is fished more like a wet fly, allowed to sink and then swing upward to imitate the hatching insect. Fact is, many wet flies are taken for caddis pupae, and not for spent mayflies or nymphs, like everybody thinks."

He continued; I could see he was enjoying the lesson as much as I was. He'd had some experience with this caddis.

"The big difference comes when you're fishing the adults, like the ones you see bouncing off the stream out there. A conventional presentation will take some fish, but not as much as if you made your fly behave in a very natural manner. A caddis doesn't just sit on the water waiting for its wings to dry. It hops, skips, and flutters trying to get off the moment it has shed its pupal shuck. They're like a thief who attracts attention by running. All that commotion arouses the interest of the trout. In a way, their quickness is their own downfall. Come on, let's see if we can't practice what I've been preaching."

And with that he stood up, stretched, and began walking into the stream. I followed eagerly.

We stood quietly in the water, side by side. He suggested I tie on a hairwing fly I'd tied up out of curiosity last winter.

"There!" he pointed.

I saw the double rings of what had been a fairly splashy rise slightly upstream and across. I cast. The fly drifted directly over the trout's window. Nothing. I cast again . . . and again . . . and again. Still nothing. The float had seemed good each time. I turned my head and looked at my instructor. He was smiling.

"Sometimes the excitement of a rising fish makes us forget all we've learned. Come with me," he said.

I followed him back out of the water, upstream for about thirty feet and back into the water. We were now in a position upstream and only slightly at an angle to the area in which the trout had been feeding. We waited.

"Now drop your fly directly in front of him, about three feet, but in his feeding lane," he said.

I did so. Still nothing.

"Make the same cast again and twitch your rod just before the fly gets to him. The fly should move upstream just an inch or two . . . no more."

The water exploded. In my excitement I set the hook too hard and the fish was off. *Son of a* . . .

The old angler laughed. "Kind of wild, aren't they? It's mostly that way with caddis fishing. No quiet slurp. No gentle take. They figure that thing is getting away and they better grab it while the grabbing's good."

"That's really something," I said.

"It's really only part of it. Took me a good long while to learn and yet it comes down simply to this: Knowing the caddis—or any other insect for that matter. If you know what they're going to do and when they're going to do it, you only have to imitate it—not just the way it looks but the way it acts."

We fished and talked most of the afternoon. What he knew, he shared with me. What he didn't know, or perhaps had forgotten, I was determined to learn. His experience and knowledge were exceeded only by his friendship.

The late afternoon sun found us still fishing the water, though I had now worked further downstream than my companion. At the moment I was into a good fish, probably the best of the day, and it took me a while to bring him to the net. Pushing sixteen inches; a beautiful brown. I turned upstream to show my prize to my new best friend, but he was gone.

I was disappointed. But now, whenever I think about it, I know that's the way he had wanted it: No maudlin good-byes; no demonstrative "Thank you's." Our meeting, though brief, was too special for such mundane manifestations. I never learned his name, nor he mine. And yet, we were angling brothers that afternoon. I'm still not quite sure if he saw me take that last good fish. I suspect he did and made his exit while I was still playing it.

Strangely, when I tell other anglers of the incident, I find that many have had a similar experience at one time or another. It supports the belief that we don't become fly fishermen by ourselves. Somewhere, sometime, someone along the way has taken us by the hand and guided us on to something new and exciting in our angling life. I believe my angler is a familiar figure to all of us.

Many days and seasons have since passed since that afternoon. Much of this time has been spent learning more about the caddis on my own. And every time I think I've discovered something new, I think back to this meeting and realize it was really something he had imparted to me, whether it was actually said in so many words, or not. That friendly lesson was the beginning of a new fishing fascination for me, one that has led to many hours of stream study and more hours tying and reading—and a good deal of plain angling fun during the season.

If, in all modesty, *The Caddis and the Angler* succeeds in this way in adding to your fly fishing enjoyment, it will have accomplished its purpose in our minds.

A final note: although *The Caddis and the Angler* is the work of two authors, including the help of many friends and contributors, it is written in the first person singular. It seemed ridiculous for us to say "We fished here", or "We tried this method". In fact, many of our experiences as related in this book occurred separately over the course of many years. To present them in a book form, nonetheless, a single "voice" seemed to be the best solution, and happily, it falls very naturally into the context of *The Caddis and the Angler.*

the caddis
and the angler

If you are fishing a lake, a stream, or a river and you see what your eyes tell you is most likely a moth, chances are that you are observing a caddis—especially if the "moth" appears during daylight hours. There is, in fact, a resemblance between moths and caddis flies in their blurred, erratic flight. Moths, however, belong to the insect order called Lepidoptera. The wings of these insects (this order includes butterflies) are covered with tiny scales. Caddis flies belong to the order Trichoptera; *trichos* is the Greek word meaning "hair"—the wings of adult caddis flies are covered with tiny hairs.

In actuality, since they appear over trout water, all across North America, caddis are more likely to be confused in the angler's mind with the better-known mayflies. And for the fly fisherman, that is really the starting place for becoming familiar with caddis: learning to distinguish them from mayflies and other aquatic insects. This is easy enough to do on the stream. Most obviously, caddis have an erratic, bouncing flight pattern. They also have *two* sets of full-sized, functioning wings (the mayflies' second set of wings is usually much smaller than the forewings), which slant backwards over the body when not in flight. This in itself distinguishes them from mayflies, which generally rest with their wings in an upright position.

But there is another level at which the angler must learn to "distinguish" the caddis from all its stream relatives, if he is to be really successful: insect behavior. After all, the only purpose we have for learning to identify stream insects is to learn how best to imitate them and make trout believe our flies are the real thing. This is especially important when we are considering the caddis, which has a natural history—including a close relationship with the trout—that is unlike the mayfly's, or that of any other insect. The caddis is, for better or worse for the angler, unique among the aquatic insects that are important to trout.

So to back up a step, the real starting place for understanding how to fish caddis imitations effectively is in our attitude towards fly fishing itself—the subtle ways in which most of us approach angling that are part cherished tradition and part force of habit. Much of our fly fishing technique has been developed around the mayflies, and for caddis, many of these traditions and attitudes must be "unlearned," or at least set aside, so that the angler may become observant of *actual stream events*. This attitude above all others pays dividends when dealing with the caddis and fishing its imitations.

Of course, observation is the key to success in all fly fishing, but it is especially important for fishing caddis imitations, where there are subtle differences between not only the caddis and mayflies, but among the caddis families themselves. In fishing with caddis imitations, the standard practices of fly fishing, especially dry fly fishing, can become blinders that prevent the angler from seeing new things—or old ones in new ways.

In fact, fly fishing's traditional emphasis on the mayflies has in some ways tended to obscure the importance of caddis to the trout and to trout fishing. To be sure, caddis do not have the delicate beauty and other admirable qualities (like regular emergences) that have made the mayflies the angler's pet. Moreover, the caddis is different from the mayfly in every way that is important to the angler: shape, behavior, and occurrence. But the fish have made their adaptation (caddis account for a full fifty percent of the trout's diet) and so must we.

All this is not to say that fishing caddis imitations is a departure from the kind of fishing we all love; it is really just another dimension, an important one for today's fisherman. Indeed, caddis imitations are certainly not new to fly fishing; perhaps without knowing it in some cases, we have been tying caddis patterns and fishing caddis imitations for some time, and there are many traditional patterns that imitate caddis in our flyboxes. But somehow caddis patterns have never assumed the significance, until recently, that anglers have accorded to the early spring mayfly hatches. Caddis patterns tended to remain localized, and even the specific insects themselves remained nameless to fly fishermen, except to the dedicated ones who were willing to attack the task of learning the Latin entomological names. There are, of course, widely known local caddis emergences with names like the shadfly and the orange sedge, but the store of knowledge and traditions available to the serious angler about caddis has never been codified in the way that it has been for mayflies. Perhaps that is because, as we shall see, the caddis is a fascinating but very complex insect.

In fishing caddis imitations, as in all fly fishing, the real link between the fish and the angler is the insect itself. And to look at the caddis as trout food we must first look at them as natural insects.

Most anglers recognize that a basic understanding of aquatic entomology—the insect's lifestyles, if you will—is essential to successful fly fishing, whether or not anglers think of themselves as amateur entomologists. And by "successful," include the less tangible benefits of a true appreciation of the streams we fish, in their own terms, and not only in fish landed.

This stream-world approach is *essential* for learning to fish caddis imitations; building a basic framework of understanding caddis behavior and life cycle into more strikes will necessarily involve some angling detective work, commonly known in scientific circles as "entomology." Here to be sure, the caddis does present some complexities: there are simply more species of caddis than mayflies (about twelve hundred), and so the caddis must be considered in its variety—or similarity—of coloration and size, the two main elements for matching the hatches. In addition, the caddis have an "extra" stage of development between the active underwater form, the larva (nymph, for the mayflies) and the winged adult, called the pupa.

Caddis actually pass through four stages of development: egg, larva, pupa, and adult—what entomologists call a complete metamorphosis. Mayflies characteristically have only three: egg, nymph, and adult. The caddis' distinctive pupa stage is really a period of hibernation during which the larva, the nymph-like form, transforms itself into an adult, much as the caterpillar does in its coccoon before it becomes a butterfly or a moth. The caddis larva accomplishes this task by building a case—an underwater home. Though the egg is not important to the angler interest in caddis, the pupa and adult stages, as we shall see, present a challenge all their own, unlike anything the angler will encounter among mayflies. Again, learning to fish these stages successfully requires a certain amount of close attention to the insect itself. Fortunately, in the case of the caddis this task has a certain fascination in itself that will naturally carry the angler along towards his aim of better fishing. It does not mean that the ardent angler who wants simply to catch more trout has to put on scuba gear, as some have done, and spend hours below watching the small underwater dramas that take place between the trout and the caddis. In most ways, entomologists have already done all the groundwork for us, and anglers need only make the connections between that information and their own tools. That is part of what *The Caddis and the Angler* is all about.

The opportunity for fishing excitement that an understanding of caddis represents begins with the bare outline of the caddis life cycle: egg, larva, pupa, and adult. But in each of these stages the caddis takes on not only a different form but a different "personality"—different behavior patterns that ultimately involve the behavior of the trout, *and* the fisherman. During the course of your involvement with fly fishing using caddis patterns, you may find, as I have, that it can become a kind of angling expedition, and the challenge of filling in the all-important details on your fishing trips becomes an aspect of fishing itself. And of course, this attitude has a counterpart in tying flies, too. But let us start filling in the details here, with the caddis natural.

The caddis fly is a remarkable insect. It is not endowed with the ephemeral beauty of the mayfly, but neither is it as fragile. A caddis is an insect with character. Sturdy and enduring, the species of Trichoptera number over a thousand in this country alone. They are found in waters in every state, emerging throughout the season. Caddis inhabit waters in which the mayfly cannot survive, whether because of acidity, high altitude, or the by-products of man's encroachment, pollution. Caddis can withstand a wider range of water temperature fluctuation. And most importantly to us, they represent over fifty percent of the food taken by trout. All these facts notwithstanding, in most anglers' minds caddis have been relegated to a fuzzy second place. All this is changing as anglers learn more and more about caddis, in a proper proportion to our knowledge of the mayflies.

Both the mayfly and the caddis comprise a major portion of the trout's diet. Each insect is taken in varying forms. The mayfly is available in three stages; nymph, dun, and spinner. The caddis offers the trout one additional stage: larva, pupa, adult, and ovipositing, or egg-laying adult. The pupal and adult stage of the caddis actually offer more excitement for the angler due to the tantalizing (to trout) behavior of the caddis during this period of his life cycle, but it is the larva that is most interesting in itself.

Though both mayflies and caddis enter a period of development underwater, that of the caddis is much shorter. The caddis larva is a worker. The caddis is also an architect. Most of us are familiar with the casemakers, those families (not all) of caddis flies that build protective cases, which are commonly found under rocks. Cases can be made of bits of twigs and vegetation such as in the genus *Brachycentrus,* which are known for their chimney type of structures; or they may be made entirely of pebbles; or a combination of pebbles and twigs or stray objects found on the stream bottom. All the larvae emit a gelatinous substance from their salivary glands which is used to glue the pieces of their case together, in addition to attaching the case to a rock or other submerged object. Each family has its own style of architecture and preference of materials used. The casemaker groups begin work immediately after entering the larval stage. The case, whatever its architectural style, becomes the home of the larva until it is time for pupation.

As the larva feeds, it grows. And as it outgrows its home, or case, it builds a new one, or expands the old one in order to accommodate its increased size.

Caddis larvae are particular about the placement of their cases on the stream bottom. The site must have protection not only from the trout and other fishes, but from the current as well.

Certain families of caddis do not build a case. One of them, Hydropsychidae, collects food and provides protection for itself by spinning nets. These nets are usually anchored between submerged rocks and act as a funnel. The larva nestles in the downstream portion while the current flows through the net, supplying food.

Another type, the genus *Rhyacophila,* neither spins a net nor builds a case. It is a completely free-living caddis, at least until it is time for pupation. At that point all larvae seal off their case, and enter a period of dormancy.

During pupation the insect goes through a transformation. It no longer eats or grows. A gross physical change takes place; the larva becomes a fully developed insect, in pupal form. When the time is ripe to make its emergence, the pupa wriggles out of its case and propels its way to the surface. The genera of most families perform the emergence ascent very rapidly, though there are those, such as in the family Limnephilidae, which crawl or swim out of the water onto rocks and branches.

(3) adult

(2) pupa emerging

free larva

(1)

cased larva

egg mass

(4)a *female/eggs*

life cycle of the caddis

Unlike both the mayfly and the stonefly, the caddis is available to the trout as a food in four stages: (1) as a larva, whether in or out of a case, (2) as a pupa, (3) as an adult and (4) as an adult female in the ovipositing stage.

(4)b *female/eggs*

The adult stage is the one we readily see flying above the surface of the stream, or gathered in colonies in trees and bushes bordering the stream. Their flight is very erratic and they are much more difficult to capture than the mayfly. Should you decide to do any collecting of adult specimens, I strongly urge you to use a net.

Mating takes place sometime after emergence, usually over the land. Unlike the mayfly, however, the adult caddis does not expire after the mating ritual. In many cases, caddis flies will mate two and even three times. Their life expectancy is approximately ten to twenty days or more, depending on the species and climatic conditions—a long full life by mayfly standards.

Also unlike the mayfly, which having laid her eggs on the surface of the stream, lies spent and exhausted after the performance, a female caddis will lay its eggs on or *below* the surface. Certain species actually dive to the bottom in order to attach their eggs to submerged rocks and return from the water flying strongly. They are, of course, extremely vulnerable to trout during the ovipositing stage, and as we will see, this stage is in some ways the most interesting to fish.

There is the basic life cycle of the caddis. Each stage has its own special challenge for the fly fisherman, and of course each stream has its own schedule of caddis emergences. Here again caddis differ from mayflies in being far less predictable. To that add the fact that trout can be extremely selective to many caddis species with similar coloration, and the full scope of the challenge emerges. The angler must first determine *which* caddis families are present, and be prepared to imitate each stage of those particular insects in their characteristic behavior. In most cases, a family-level identification will suffice, especially for the larva and pupa; for the adult a genus or species identification may be necessary. Again, fishing caddis imitations is most often a stream-by-stream affair of finding and identifying, tying and fishing.

So the process of becoming familiar with caddis hatches in local streams is in many ways a frontier. For the fly fisherman who enjoys this element in his fishing days, each new stretch of water is an opportunity for discovery. *The Caddis and the Angler* is designed to help in that task.

Because caddis flies have for many years been carried in anglers' fly boxes as a second choice—a "just in case" substitute should the mayfly strike out—there are very few widely known common names for caddis.

Yet, in order to fish them properly you will have to become familiar with them, one step at a time. Therein lies the secret to knowing them all, "one at a time." The pages that follow will list the proper scientific names for the various families and genera of caddis flies. It is not as important to know all of the Latin names as it is to know the habits and characteristics of certain groups of caddis in general. If you try to absorb all of it all at once you will only follow a confusion of paths. In fact, it is best not to be concerned at all in the beginning. You will find that as you become familiar with certain steps, little bits of information will begin to cling here and there. The pieces of the puzzle will gradually fall in, and gradually you will have compiled a fountain of knowledge you did not believe probable when you first began. You will find it useful to refer to the "Caddis Families" section of this chapter as you read and tie, to begin building this familiarity. These profiles will help give the family names essential meaning, and using Caddis Families as a quick reference will answer most questions you may have about caddis specimens.

Let's look at the caddis family by family.

habits and characteristics

caddis families

The order of Trichoptera consists of approximately eighteen families in the North American continent and about thirty throughout the world, including about twelve hundred species. However, not all of them are important to the angler; only those families which have proven to be a consistent food supply for the trout, and thus a reason for the angler to tie and fish them, have been listed.

The size of the wings, legs, and bodies will vary within the respective families, as will the color. Sizes given are an average, or approximate to those most commonly collected. Unless otherwise specified, the color shading will be descriptive of the adult stage of the caddis.

Should you decide to do your own collecting, you will find that the sizes and colors given in the following family profiles will correspond with your own findings, for the most part. Some notable exceptions will be found in the families of Leptoceridae and Limnephilidae. Both of these families are so varied that it is difficult to arrive at an average.

For the sake of accuracy the size of the insect will be referred to in millimeters (mm). Again, the sizes are for the adult insect. As a general rule, especially with Ryacophilidae and Hydropsychidae (which are of prime importance for larval imitations since they are free-living in the larva stage), the size of the larva is approximately twenty to twenty-five percent larger than the full length of the adult from head to wing tip. The main purpose of these profiles is to help anglers build a familiarity with caddis families and their characteristics; for tying purposes, consult the insect to hook size correlation chart in Chapter 4.

brachycentridae

habitat   Found in the riffles of rivers and streams throughout most of the United States and Canada (Southwest excepted).

case or home   Case consists of cross sections of vegetation formed into the shape of a chimney. (On occasion the case may be only of silk.) The front of the case is attached to a rock, while the legs and head protrude from the open end, searching for food.

emergence   From spring to early summer, often in great quantities. Ascent is rapid.

color   Adult is brownish/olive to greyish/yellow. Larvae are greyish/olive to greyish/yellow.

ovipositing   Eggs are laid from massed ball. Female either dips her abdomen on the surface of the stream, or actually enters the water and secures the eggs to submerged rocks.

wing shape

average adult   size: Adult varies 10 to 14 mm from head to tip of wing.
legs: Rusty dun to bronze dun.
wing: Tannish grey to mottled greyish brown.
body: Greyish yellow to greyish olive (dark).

A small family, now part of Limnephilidae.

goeridae

habitat   Found in fast-flowing water. Common throughout the United States, but mostly an eastern variety.

case or home   Case is a short tub made of pebbles with larger stones cemented to sides to give a winglike appearance.

emergence   Early spring through summer. It emerges once the pupa has crawled out of the water.

color   Adults are mainly greyish brown. Larvae are tan or brownish.

ovipositing   Egg masses are deposited in the water and also damp areas nearby.

wing shape

average adult   size: 5 to 7 mm.
legs: Rusty dun to bronze dun.
wing: Greyish brown.
body: Greyish brown.

helicopsychidae

Only four known species in the United States, found primarily in the Southwest. However, *borealis* is quite abundant and widespread across the country and up to southern Canada.

habitat	Prefers flowing water; fast and medium riffles.
case or home	A snail-type case made of pebbles or stones.

emergence	Emerges sporadically throughout the season
ovipositing	Deposits egg masses in or near the water. If near the water it is in a damp area.
color	Brownish grey to tannish grey adults. Larvae are brownish.
wing shape	

average adult size: 5 to 7 mm.
legs: Rusty dun.
wing: Brownish grey.
body: Tannish grey to brownish grey.

Most common of all caddis.

hydropsychidae

habitat   Found throughout the United States. Prefers flowing water with rocky bottom. Freestone streams offer excellent habitat.

case or home   Larvae build a silken retreat in rocks. Web is extended to form a seine net, thus capturing food as a current funnels through. At pupation the web is enlarged to a silken dome-shaped case, often composed of grains of sand.

emergence   Very rapid through both water and surface. Occurs during early spring till summer.

color   Straw to greyish/olive to brown adult. The larva is greyish/green to off white; varies greatly.

ovipositing   Female crawls into water and lays her eggs on the bottom, and is known to emerge and mate again after ovipositing.

wing shape

13MM = 1/2"

average adult   size: 10 to 12 mm from head to wing tip.
legs: Light medium rusty dun.
wing: Mottled medium brownish grey.
body: Greyish olive to brownish grey.

notes   There will usually be at least six to eight species present in a particular stretch of stream. Emergence will vary with times of year. Adults of different species are very similar in appearance.

lepidostomatidae

habitat   This family is widespread throughout the United States, covering approximately three dozen species. Found mostly in streams or springs but occasionally in cold water ponds. Prefers quieter water, such as pools.

case or home   Cone-shaped cases (like *Brachycentrus*) are made of sand and vegetation. They may be round or square in cross section. Some species build log cabin type or round stone cases.

emergence   Mainly spring.

color   Tannish/grey to dark grey. Larvae are mainly cream to brown.

ovipositing   Egg masses are dropped on damp areas near the water, or actually in the water.

wing shape

average adult   size: 8.5 to 9.5 mm.
legs: Rusty dun to dark grey.
wing: Brownish grey to grey.
body: Tannish grey to dark grey.

habitat	Found in both fast and still waters throughout most of the United States. Fairly common.	**leptoceridae**
case or home	Cases vary from straight to curved. Some are conical and others cylindrical. Materials also vary from grains of sand to twigs and vegetation.	

emergence	Mainly early summer.
color	Adults vary from straw to yellow to brown. Larvae are brownish yellow.
ovipositing	Balled egg masses are dipped into the water by the female. Upon sinking, eggs attach themselves to objects on the stream bottom.
wing shape	

average adult	size: 8 to 10 mm. legs: Straw to light brown. wing: Cream to light brown with darker markings. body: Yellow to brown.
note	Family has very long antennae, but there is a great variation among species.

One of the largest of families.

limnephilidae

habitat   Found mostly in the northern sections of the United States and the higher elevations of the Southwest. Prefers slow sections of streams in addition to pools, lakes and bogs. However, there are some species, such as those in genus *Neophylax,* that inhabit riffles.

case or home

Materials used for the case vary greatly throughout the family though the individual species are fairly selective. Generally, the cases are large affairs constructed of sticks or stones, or both.

emergence   The major portion of emergence takes place during late summer and autumn. However, in the northeastern states and southeastern Canada, emergence usually occurs during late spring and early summer. Again, there are exceptions in this family. Unlike other families, the pupa of Limnephilidae crawls or swims out of the water and changes to adult form on land.

color   Straw to reddish/brown. Larvae color is cream to brown.

ovipositing   Females display an egg sac; eggs are laid along the sides of the stream in moss or damp areas. After hatching, the larva crawls or is washed into the stream or body of water. (There is one species in Europe, *enoicya,* that lives its entire life cycle in damp areas on the land.)

wing shape

average adult   size: (Only approximate due to variation) 15 to 20 mm.
legs: Medium brown.
wing: Medium brown and mottled.
legs: Light brown.

notes   The species are quite varied in habits. There are over twenty genera comprising nearly two hundred species. Most species are about 14 to 19 mm in size though there are a few significant variables.

A small family known for the genus *Psilotreta*. This caddis is sealed in its case at the end of its larval, as well as during its pupal stage. *Psilotreta* is almost exclusively eastern.

odonticeridae

habitat	Prefers flowing water in rivers and streams.
case or home	Fine grains of sand are used to build a tube-shaped case. The cases are usually massed together and attached to sides of rocks in clusters *(Psilotreta labida)*.
emergence	Water temperature (around 60°) sets the time for emergence. Failing light is another activator. Late spring weather triggers the period of emergence, which is narrowly restricted. Early emergers have a shorter life span than the late arrivals. Males precede the females by a day, or as much as a week. The pupa is very fast in its burst for the surface.
color	Adults are dark grey with green/black bodies. Larvae are primarily green.
ovipositing	Egg masses are deposited on the surface. After settling, the eggs attach themselves to submerged objects.

wing shape

average adult size: 14 to 16 mm.
legs: Bronze dun to grey.
wing: Brownish grey to charcoal grey.
body: Greenish black to charcoal grey.

philopotamidae

habitat
Found more profusely in the eastern states but relatively abundant throughout. Comfortable in flowing waters, from tiny springs to large rivers.

case or home
This extremely active family weaves silken nets attached to the undersides of rocks. Referred to as "finger nets," they are long and narrow while the current flows through them, but these structures collapse into a mass of silken folds when they are removed from the water. During pupation, the form of construction changes. A dome-shaped case of tiny pebbles is formed and attached to the underside of a rock or other object.

emergence
A fairly rapid emerger. The genus *Chimarra* has its peak emergence early in the season, with some late exceptions, while *Dolophilodes* can be found hatching almost all year long (in the winter, females have been found wingless, crawling on the snow of a stream bank during emergence).

color
Brown to almost black. Larvae are cream to almost orange.

ovipositing
Females enter the water and lay their eggs in strings on rocks.

wing shape

average adult
size: 6 to 8 mm; larvae are 10 to 12 mm.
legs: Rusty dun to dark bronze dun.
wing: Greyish brown.
body: Brownish grey to charcoal grey.

note
Antennae are usually longer than the body.

This family is noted for its large cases which usually measure from one to two inches. Adult life span can be up to several months.

phryganeidae

habitat   Northerly United States, both East and West. Prefers the higher elevations. Likes still water such as pools of rivers, lakes, and ponds.

case or home   Large (65 to 70 mm), slender, and spiral-shaped, the cases are constructed from vegetation. Larvae readily leave their cases during stress.

emergence   Peak emergence is spring and early summer. Members of the family crawl out of the stream as pupae and emerge.

color   Adult varies from yellow/brown to dark brown. Larvae from yellowish/brown to medium brown.

ovipositing   Eggs are deposited in masses beneath the surface on sticks or other material. Usually an "egg wreath" attached to a stick.

wing shape

average adult   size: 18 to 25 mm.
                             legs: Mottled ginger to medium brown.
                             wing: Mottled tan to medium brown.
                             body: Yellowish brown to brown.

polycentropidae (psychomyiidae)

Polycentropidae and Psychomyiidae were formerly classified as one family. Their habits and basic characteristics are similar. However, since Psychomyiidae average only 3 to 5 mm in length, we will not discuss them separately since the larger Polycentropodae may be more important to the angler.

They are distributed all across the country.

habitat	They inhabit mainly small and large streams; however, they may be found in northern ponds.
case or home	They spin a netlike retreat, often giving a trumpet-shaped appearance. At pupation they form a silken case and afterwards reinforce it with fine sand grains or vegetation.
emergence	Like the Hydropsychidae, they may emerge sporadically throughout the season, and their ascent from the bottom is usually rapid.
colors	Adults are various shades of brown with light, almost silvery flecking on the wings. Larvae are straw-colored.
ovipositing	Enter the water and lay strings of eggs on the bottom.

wing shape

average adult size: 7 to 9 mm.
legs: Light brown to rusty dun.
wings: Various shades of brown with lighter flecking or mottling.
body: Light brown to greyish brown.

One of the most widely distributed families. One genus, *Glossosoma,* has been given the status of family (Glossosomatidae), but since the characteristics and behavior of this group remain very similar to the other Rhyacophilidae, it will not be discussed separately here.

rhyacophilidae (glossosomatidae)

habitat   Common across North America. Prefers fast water, especially in small streams.

case or home   This family has free-living larvae about 10 to 20 mm. There are no cases until pupation. During the pupal stage a dome of pebbles is formed which is attached to a solid object. A silken coccoon is formed within the domed case. An important exception is *Glossosoma* which builds the pebble case during the larval stage.

emergence   Heaviest activity is during late spring and early summer, though emergence occurs throughout the year. Rise to the surface is fairly rapid.

color   Adults are dark tan to greyish brown. Larvae are various shades of green tending to brown.

ovipositing   Female propels herself into the water and deposits her eggs in sheets along the bottom.

wing shape

average adult   size: 8 to 13 mm.
legs: Tannish grey to brownish grey.
wing: Grey/brown mottled to dark grey.
body: Brown to brownish grey.

the time
of the
larva

after the adult female caddis deposits her eggs, a gelluloid substance with which nature has endowed these eggs allows them to cling to submerged rocks. In due course of time, usually a matter of a few weeks depending on the species, the eggs hatch and the first form of caddis life important to the angler, known as the larva, emerges. Outside of the approximate two or three weeks spent as a pupa, and two to eight as a winged adult, the caddis will spend most of his life cycle in this form, which is quite interesting in itself.

Since the grub-like larvae have no defense against predators, many caddis species—but *not* all—immediately go about the task of building a kind of shell, or case of various pieces of twigs or wood fragments, or sand pebbles and twigs, or perhaps just pebbles. The methods of construction differ with the various families, genus, and sometimes species (see pages 26–37). The cases offer excellent camouflage for the larvae, and will later serve as a protective shelter when they mature to the next phase of development, the pupa.

The larval cases are small wonders of natural engineering—the miniature chimneys of *Brachycentrus,* a very common caddis casemaker, are remarkably symmetrical. Other cases seem to make almost random use of whatever materials are available to the larva. But in fact all of them are adapted to the stream environment of the species, including most notably the speed of current flow. Casemakers in lakes and ponds, for example, make the most loosely constructed cases. Cases found in fast water are usually very solid, and more likely made of pebbles than other materials.

In the stream world what seems obvious or inconsequential to the angler is often the key to survival. If you look closely at caddis cases found under stream rocks, you may wonder why certain designs include long sticks that protrude beyond the length of the actual case, or why extra bits of pebbles are attached to the sides of the cases, like wings on a plane. These seemingly eccentric bits of structure do not appear to have any purpose—and yet they do. They serve as ballast in the fast-flowing waters of the stream and they discourage some of the smaller-but-hungry fish from trying to devour a bit of food that is just too large for them to handle.

The casemaker larvae accomplish this underwater construction project using well-developed mouth parts and a remarkable jelly-like substance that comes from their salivary glands and acts like a super-glue, holding together the bits of building material which they grab as it floats by, or find on the stream bottom. In this way they gradually envelop themselves with the case, and as they grow the larvae will expand their case and lengthen it. Some species will simply add more material to their case; others will build new and bigger

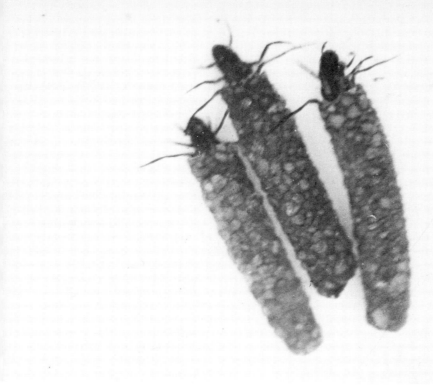

The *Psilotreta labida* larva makes its case entirely of small pebbles; they are usually found clustered together in colonies under stream rocks.

A *Dicosmeocus atripes* larva, one of the largest caddis species, in its stone case. At hatching time the *Dicosmeocus* species crawl or swim along the bottom to shoreline rocks, where they will burst the pupal shuck.

quarters. In either "case," at this point they become more vulnerable to trout, which is significant for the fly fisherman standing in the stream around them.

Casemaker larvae are relatively mobile (although one genus of casemakers, *Glossosoma,* settles down permanently on a protective rock), dragging their case with them around their rock or stream vegetation habitat as they look for food in the form of tiny bits of aquatic life: diatoms, plankton, other insect larvae, including on occasion, each other.

Although the casemakers are a fascinating form of stream life, other caddis families have developed adaptations for survival that are even more important to the angler.

Some families, such as Hydropsychidae and Philopotamidae, build silky netlike retreats with a spiderweb kind of filament they excrete from their glands. Unlike their casemaking cousins, these larvae usually fix their nets to the stream bottom in such a way that the net can shelter the larvae and filter food from the stream flow at the same time—between crevices in rocks, for example. They can move in and out at will, and the nets always have an escape exit at the rear.

Still other caddis families, such as Rhyacophila and Hydroptilidae are free-living; they do not construct any case at all during their larval period. Larvae of these species have their own adaptations for this homeless lifestyle in strong legs and anal hooks to help them grip the bottom. The free-living forms build a case only when they are ready to pupate.

There are other adaptations, such as the tiny tube or burrowing caddis, but these last two groups, the netmakers and free-living caddis, are by far the most important caddis for the angler, the ones we usually have in mind when tying larva imitations.

Author's note: In the first chapter of this volume comment was made concerning the hardiness of caddis as opposed to the mayfly. Gary LaFontaine in his *The Challenge of the Trout,* disagrees. He claims that both are equally hardy, and the reason for the increase or seeming preponderance of caddis flies in our streams is due to the increase of organic leachings of civilization, giving many filter-feeding Trichoptera species, the netmakers, a competitive advantage over mayflies. Many netmaking caddis, such as *Hydropsyche,* are more abundant because nutrient-laden currents now deliver greater quantities of drift food into their feeding nets. This is an interesting observation and I for one am glad to hear that our tampering with nature has had at least some benefits for aquatic insects and trout.

Limnephilidae are typical casemakers, using sticks and stones in a variety of designs, although each species' cases, like *Pycnopsyche guttifer* here, will usually be distinctive.

Caddis larvae of all kinds can be found throughout the stream habitats: over and under rocks, and in vegetation. Trout and other fish feed on most larval forms whether they are in a case, a net, or living freely. The trout poke around the rocks and root out cases and ingest them, twigs and all. The larva itself will be ingested as food, while the wood fragments and twigs will be excreted. But most often, except for the very minute early stages of caddis, trout will see very few caddis larvae that are uncased. It makes you wonder why we bother tying this type of imitation at all.

To answer that question for yourself, here is an experiment you may wish to try (only for scientific purposes, since after all, you are a fly fisherman). If you were to take a caddis case containing a live larva, and impale it on a hook, you might catch a certain number of trout. If, however, you remove the larva from the case and use the larva alone as bait, your results would improve significantly.

Likewise, an imitation of a caddis case (which has been done) is not as effective as a larva imitation. Although some fish will take a case imitation, tying them is a waste of time in relation to the rewards received as strikes.

And most casemaker larvae spend their time either attached to a rock or hiding in submerged vegetation. That fact limits their value to the fly fisherman. About the only way to imitate their natural activity is to snag your fly on the bottom, and wait for a trout to come along looking for a cased larva of that species. However, the several families in which the larva is not cased are very noticeable to trout and other fish, since they move more freely in their habitat area. These are the ones worth imitating and fishing.

In addition aquatic larvae of all orders are subject to a certain phenomenon called "stream drift" or "catastrophic drift." It is a random breaking away of larvae or nymphs from submerged rocks and objects to drift aimlessly for some distance before again attaching themselves to another stream-bottom position. Since it is especially likely when conditions in a particular part of the stream or even a particular rock become overcrowded, perhaps stream drift is nature's way of preventing overpopulation (in these conditions, caddis larvae often become cannibals).

Although this phenomenon has been known to entomologists for some time, it's hard to apply it to angling, because it seems to be almost unpredictable. The most important factor which triggers this occurrence is light—or rather the lack of it. In an experiment conducted by Gary LaFontaine of Deer Lodge, Montana, the following stream drift data was recorded.

1) Cased caddis larvae reached their peak incidence of stream drift at 7:30 A.M. and 7:30 P.M.
2) Mayfly nymphs, stonefly nymphs, and uncased caddis larvae reached peak incidence in Gary's sampling nets at 9:00 P.M. and 4:30 A.M. (a minor peak was reached at 1:15 A.M.)
3) Midge larvae were represented in an even distribution during a twenty-four hour period.

On moonlit nights the incidence of stream drift in Gary's experiments was less than on nights with no moon; but otherwise, there seems to be no pattern to this tantalizing behavior. The larvae may move only a short distance to their new territory, but during this time they are very vulnerable to fish—a good reason to take up night fishing.

In order to know how and with what to fish we must first determine what the natural itself looks like. To most of us the larva will appear to be a tiny worm or grub, approximately three-quarters of an inch long. Upon closer inspection, however, you will notice that the body is made up of segmentations more pronounced in some species than others. The larva will always appear to be in a curved state with the convex surface forming the back and the concave the belly. If you uncurl it and look at the underside, you will also notice tiny gill fibers protruding. The prominence of the larval gills varies from family to family; the free-living forms tend to be more lightly gilled, but this is no hard and fast rule. Species of Hydropsychidae have very pronounced gills, while in Rhyacophilidae they are insignificant. At the same time, the body segmentation of the free-living Rhyacophilidae is much more pronounced.

The gills on larvae are fairly short, not unlike the legs which are found on shrimps or scud. These gills develop singly or in clusters, depending on the species, but most importantly to the fly tyer, they weave and move as the water passes through them.

A *Hydropsyche* larva exhibits the heavy gilling common among the netmakers.

A typical *Hydropsyche* larva; the Hydropsychidae
is composed of netmakers, and characterized by
heavy gilling. With Rhyacophilidae, they are one of
the two most important caddis larvae.

The free-living *Rhyacophila fusula* is representative
of the Rhyacophilidae, important to anglers since it
is a noncasemaker larva. Notice the lack of gills and
the pronounced body segmentation. The common
fuscula is usually bright green.

Caddis larvae do have a head which is usually darker than the rest of the body. Behind the head are three pairs of semi-formed legs, which also move.

Body color of the various species varies from cream to brown, and light green to olive or grayish olive. Different shades of green seem to dominate the species that are readily accessible to trout in larval form, that is, the netmakers and free-living families.

As far as the fishing and tying of this stage is concerned, there is not much to learn from the natural. It is the easiest and least complicated of the four stages.

fishing the larva

Larva imitations can be fished almost year round, since there is almost always a "class" of larva-stage caddis in a stream at any given time; however, larva imitations work best when little or no surface activity is taking place, and trout are foraging along the bottom for food—perhaps because they know caddis larvae are readily accessible, even during lean times.

Unless the water to be fished is exceptionally deep, a conventional floating line may be used. An uncased larva that has been caught by the whim of the current will generally be free-floating. However, it is an asset to keep your fly as close to the bottom as possible, since any dislodged larva will most likely be tumbling along deep, right over the rocks. And, of course, trout are usually concentrating on the bottom for tidbits if there is no significant surface activity. Simply cast across, or slightly upstream and across and let your fly and leader sink on a slack line. Watch the end of your fly line closely for the slightest indication of a take. A line indicator may be helpful.

Allen Breault, of Fitchburg, Massachusetts, informed me that he uses a sinking line (an exception to the floating line advisement), thus allowing the larva imitation to roll and bounce along the stream bottom with exceptionally good results. On the other hand, some anglers prefer to weight their patterns or use tiny split shot to obtain the same effect, that is, getting the fly on or near the bottom.

A larva imitation can also be walked. This is accomplished by stripping forty to fifty feet of line from the reel, making a thirty-foot S cast, and feeding line through the guides in order to keep the imitation working downstream near the bottom in a dead drift. An entire section of streambed can be covered in this manner by changing positions on a line across the stream.

There are a number of ways to tie a caddis larva imitation; sor
new. The one currently most popular is the Latex Caddis Lar
Raleigh Boaze, Jr., which creates an unusually smooth-bodied i

The old reliable Breadcrust uses a traditional wool or floss b
proaches it is important to imitate the moving parts of the partic
gills and legs—and its coloration. Generally, the larger the larva t
has. And remember to match the larval coloration to the *wet* col
terial (this will also apply to pupa ties). You will find a key to larva
most families (larva color will vary within some families) in the Ca
section of the Appendices. Both the free-living and netmaking
listed. These are the ones most likely to be found by trout, but as
perimentation is the fly tyer's best friend.

While it is always a good policy to tie a number of sizes of each color, if you
work from the Emergence Tables in Chapter 6, or reliable local information, you
can safely tie the larva twenty to twenty-five percent size larger than the adult
size listed in the charts.

Black thread is the usually called for in the larva patterns, since almost all
larvae have a dark head, no matter which body coloration the species exhibits.

Experimentation led to Raleigh Boaze's new caddis larva patterns, which
are tied with a material called latex. Latex was actually used by the late Roy
Patrick of Seattle, Washington in the early fifties, but did not catch on at that
time. Since this rubbery material is a solid substance and it has no moving parts
when wound on a hook, the latex patterns are in some ways a contradiction to
the traditional idea of imitating the natural's motion. I've personally asked a
number of fishermen how successful they were when they used this type of
larva imitation and received both negative and positive answers. In some areas,
latex larvae are nothing short of spectacular, and their success may not be a
contradiction at all, but the result of close imitation of smooth-skinned naturals.

The larvae of Rhyacophilidae, a very widespread caddis, are very smooth-
skinned and do not display the heavy gilling characteristics of some of the other
families. Therefore, the smooth latex imitations could easily imitate its various
species.

latex caddis larva

Latex is a very fine, cream-colored sheet of rubber material. Cut into strips and
wound on a hook shank in an overlapping fashion, it gives the most realistic imitation when caddis larva patterns are being tied. Here is the pattern description
of the most commonly used Latex Caddis Larva.

latex
caddis
larva

hook: Mustad 37160, or similar English bait hook

thread: black

underbody: olive green floss

body: cream latex

head: peacock herl

1 This pattern is tied in sizes eight through sixteen. You will notice that the hook recommended has a pronounced curve to it, in order to imitate the shape of the natural.

Place a preferred hook in your vise keeping the forward shank portion on a horizontal plane. Spiral the thread onto the shank beginning behind the eye and winding to the bend.

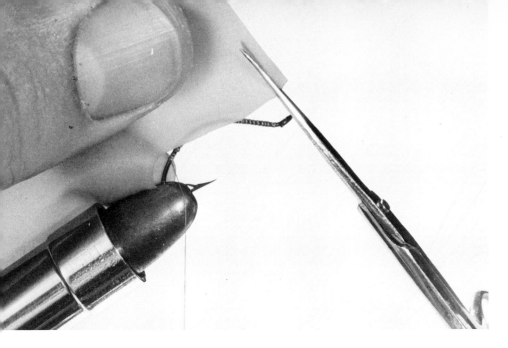

2 The next step is to cut an eighth-inch wide strip of latex approximately four inches long. (Incidentally, latex sheets come in two thicknesses. One is twice as thick as the other. Supply houses usually sell the thinnest as "regular" and the heavier gauge as "heavy duty". Both do the job, though the heavy duty form is easier to work with.)

When you cut the strip of latex from the sheet you will find that it presents a minor problem in permitting a perfectly straight cut. Don't worry if the edges are a bit uneven. It will straighten itself out when it is wrapped around the hook for a body.

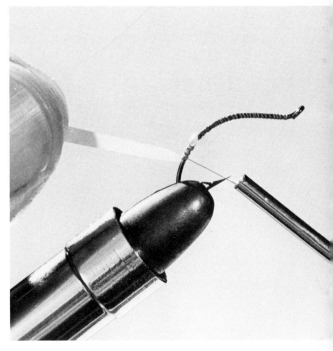

3 As you finish cutting the four-inch strip of latex taper the last portion of it to a point, or tip.

Place the latex against the hook shank at a point well down into the bend and tie it in by the tip.

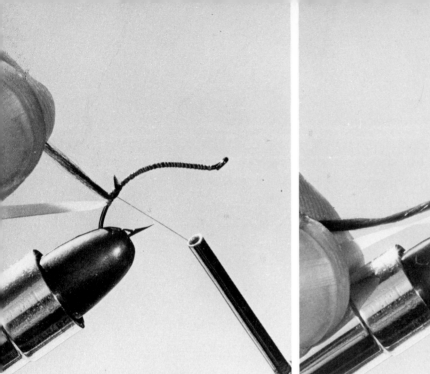

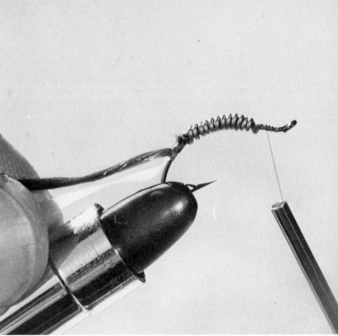

4 Cut an eight-inch section of olive green floss and tie it in just forward of the latex strip.

5 The next step is completely optional. Some tiers prefer to weight their nymph and larva imitations in order to get them down deep during high water conditions. If you are opposed to weighting your flies, the same result can be obtained simply by using a sinking line. In any event, weight can be added in the tie by winding a section of fine lead wire around the shank of the hook. A liberal amount of head cement applied to the shank will assist in keeping the lead wire in place. The tying thread is then wound between the spirals of lead wire for additional security.

Whether or not the larva is weighted, the thread should be brought forward to a position about one-eighth inch before the eye of the hook.

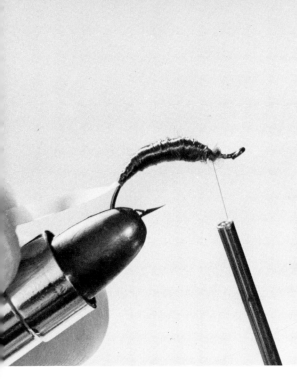

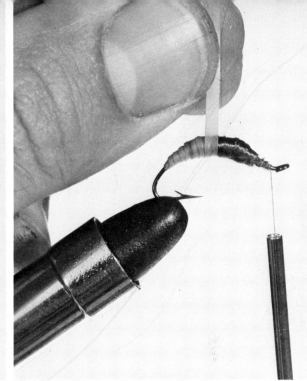

6 The floss, tied in previously, is wound to the thread in a natural taper where it is tied down and the excess clipped.

7 The strip of latex, which has been left dangling at the bend, is now wound on as the body of the fly. It is wrapped around the shank of the hook in overlapping spirals. You will find that there is a certain amount of stretch to the latex material, but it should not be pulled too taut. Only enough tension should be applied so that the latex is snug to the body and the overlap of latex forms a definite ridge, or segmentation, in the appearance of the body.

8 The strip of latex is wound to the thread, with which it is tied down, and the excess clipped. A touch of head cement to the windings at this point will give added security to the fly.

LARVA

9 Select two peacock herl fibers for the head of the pattern. The flues on the herl should be fairly long. Those fibers having the longest flue are generally found just below the "eyed" portion of a peacock tail.

Lay the peacock fibers on top of the hook shank just behind the eye and tie them in with your thread.

10 Wrap the peacock herl around the shank of the hook behind the eye and form a small but bushy head. Whip finish and cement the windings. Your completed Latex Caddis Larva should appear as the one in this photo.

Again, though it is the most common, olive-colored larva imitations are not the only ones called for. Other larva colors include cream, tans, greens, and greys. Latex can be dyed to these colors with conventional dyes such as Rit or Tintex, which can be obtained in local department stores or supermarkets.

If you use the regular or fine latex material you can obtain subtle realistic effects, depending on the color of the material used for the underbody. A certain amount of that shade will show through the latex. The application of head cement to the underbody will intensify the effect.

Another trick is to use a marking pen, either for a two-toned segmentation effect (by shading only half the strip) or a solid color.

In our pattern we've used floss as an underbody. Almost any other material such as wool, poly yarn, mohair, or even fur, may be used for that purpose. You may also wish to try wrapping the strip of latex in an open spiral thereby causing some of the underbody to be exposed. If you used wool, mohair, or fur, you can obtain the effect of tiny gills protruding from the body. There are many variations in the ways a Latex Caddis Larva can be tied.

the breadcrust

The Breadcrust pattern has long been used as an all-purpose subsurface pattern and proven itself very effective. There are quite a number of anglers who would not dream of going astream without their favorite Breadcrust pattern.

Quite a few tiers dress this pattern on a standard wet fly hook such as the Mustad models 3906 or 3399. However, I believe the use of the bait style of hook more closely resembles the larva we are imitating.

breadcrust

hook:	Mustad 37160 or equivalent
thread:	black
body:	orange or green floss or wool
rib:	center stem of dark brown rooster feather
hackle:	grizzly

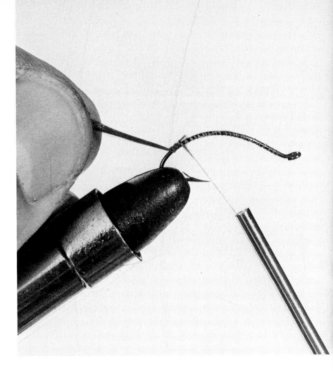

1 The Breadcrust is tied in sizes eight through eighteen. Place a hook in your vise and spiral a fine black thread onto the shank, beginning just behind the eye and terminating well into the bend.

Cut an eight-inch section of burnt orange floss or wool (or the proper color to match the larva you want to imitate) and tie it in. Let it hang back over the vise for now. If you wish to weight your fly, now is the time to do it (see the procedure described for Latex Caddis Larva).

2 From a dark brown coachman or furnace type of rooster neck, pluck one of the larger feathers near the base which has a distinct reddish shade to its center stem. Remove all the hackle fibers from the stem either by pulling them straight out, or dissolving them in clorox. *Do not strip them from the stem.* Stripping will remove the reddish pigmentation.

3 Tie the rooster quill stem in by the tip, or thin end, just above the previously tied-in floss.

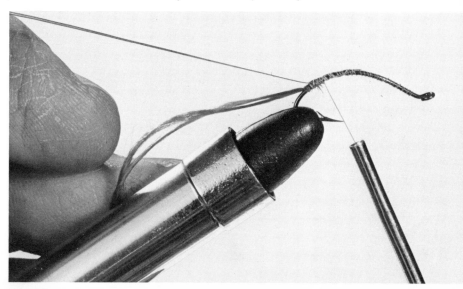

LARVA

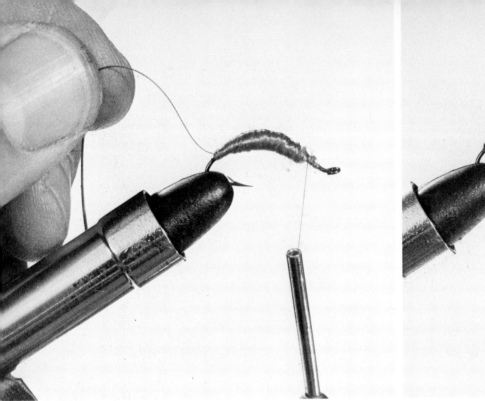
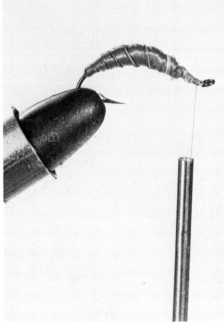

4 Bring the thread forward to a point one-eighth inch before the eye of the hook.

Grasp the floss and wind it to the thread, forming a neatly tapered body as you do so. Bind it down with the thread and clip the excess.

5 Grasp the rooster quill stem and wind it in an open spiral to the thread. It should be wound in reverse turns, coming toward you, so that it does not sink into the floss. Tie it down well and clip the excess. Apply a touch of head cement to the windings.

6 Select a soft and webby hackle fiber from a barred rock or grizzly rooster neck.

7 Trim the excess fibers from the base and tie the hackle feather in just behind the eye.

8 In connecting turns progressing toward the eye, wind the hackle around the shank of the hook. Three or four turns should be adequate.

   The required turns of hackle having been made, clip the excess butt.

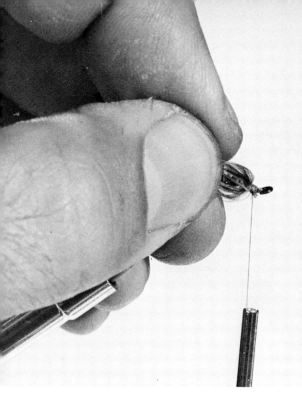

9 With the forefingers and thumb of your free hand hold the hackle fibers in a rearward position and wind turns of thread in front and partially over them, forcing the fibers to remain in a rearward slant. Whip finish and cement the head.

10 The completed Breadcrust.

gill-ribbed larva

hook: Mustad 37160 or equivalent

thread: Black

body: Green floss

rib: Peacock or ostrich herl

head: Peacock or ostrich herl

This particular pattern has a prominent gill-like effect which makes it very appropriate for imitating the *Hydropsyche* species).

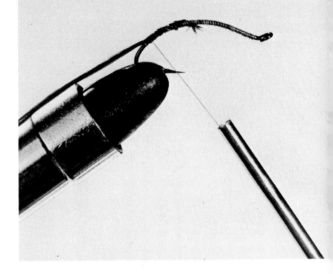

1 This pattern is tied in sizes fourteen to twenty. Place a hook in your vise and spiral your thread to the bend. Cut an eight-inch strip of green floss, and tie it in at the bend.

2 Cut a six-inch section of gold or copper, wire and tie that in on top of the floss.

3 Keep both the tied-in floss and thread out of the way for now. If you have a material clip, insert them in that. Otherwise, just slant them rearward over the vise.

Select a long-flued peacock herl fiber and tie it in by the tip just forward of the floss.

4 Bring the thread forward to a point one-eighth inch behind the eye of the hook.

Grasp the floss and wind it to the thread, forming a neatly tapered body as you do so.

LARVA

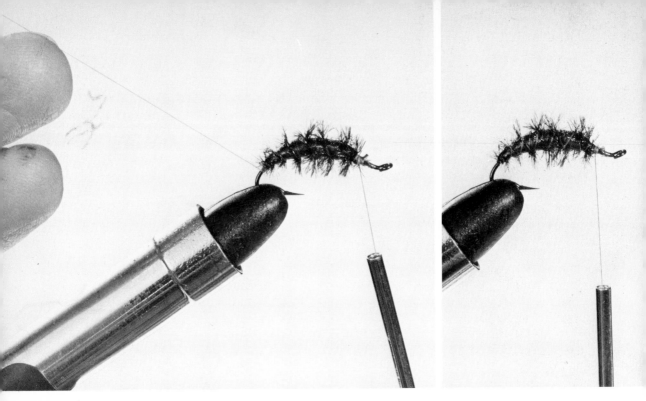

5 Grasp the peacock herl fiber and wind it through the floss body in an open spiral terminating at the thread. Tie it down with the thread and clip the excess.

6 Grasp the piece of gold wire and wind it through the body of the fly in a counter-clockwise rotation to the thread. In other words, instead of winding over the shank and away from yourself, wind it around the hook toward yourself. The gold wire is only used to lend security to the peacock herl rib, which is very fragile. Upon reaching the thread, which is hanging by bobbin tension, tie it down also and clip the excess.

larva: points to remember

1 Larva imitations work best when no surface feeding is taking place.

2 Larva imitations should be fished as close to the bottom of the stream as possible. A rolling, tumbling, at-the-mercy-of-the-current effect produces the best results.

7 Select two full-flued peacock herl fibers and tie them to the shank, just behind the eye.

8 Wrap the peacock herl around the shank of the hook just behind the eye, creating a compact and bushy head. Trim the excess, whip-finish, and add a drop of head cement. The Gill-Ribbed Larva is complete.

3 Latex-type imitations work best when species of Rhyacophilidae are present, since Latex pupa ties imitate the smooth-skinned insect. Other species have pronounced hairlike gilling characteristics, and the imitations should simulate these moving parts.

pupa
a dangerous age

When the larva of the caddis has reached its maturity, it seeks a place to attach its case or build one permanently, using the glue it used to construct the case.

The noncasemakers, such as the free-living Rhyacophilidae or the net-spinning Hydropsychidae and Philopotamidae, make cases *only* at pupation. Rhyacophilidae and the net-making Philopotamidae form separate dome-shaped cases from pebbles lined with silk and attach them to rocks and other objects, while the Hydropsychidae simply fortify their net, into which they can retreat during this transformation, with fine grains of sand.

The case is sealed up, with the exception of a few tiny holes. This ventilation system is necessary to permit a certain amount of water to pass through for the aeration of the pupal gills.

The larva no longer eats. Once the case has been sealed, a transformation takes place during which the larva becomes a pre-formed adult. Over a period of about three weeks it develops its elongated adult legs, genitalia, antennae, and a semi-formed wing inside an enveloping skin, commonly called a shuck.

The pupa often undergoes color changes. Much of the larva's common green coloration is caused by algae in its diet and when feeding stops at pupation, the strong green overtones dissipate. Such changes of coloration continue even after the emerged adult has matured. Green coloration may be retained in the adult female, caused by the eggs she carries prior to ovipositing. The pupa and the newly emerged adult, however, will be very similar in shade to each other.

When the proper conditions for emergence have occurred, the pupa breaks open the end of the larval case at the head. As soon as an adequate escape hatch has been cut, the pupa wriggles out of the case and is very quickly on its way to the surface. Just prior to emergence, the pupa collects bubbles of gases which will help separate the pupal shuck from the body of the emerging adult. These minute gas bubbles are also the main means of the rapid ascent of most species. The effect is like holding an air-filled balloon at the bottom of a pool and letting it go up like a rocket. That's why trout and other fish have to move so fast to capture them. There are exceptions. On some rare occasions, the pupa will swim to the surface relatively slowly. Many species of the Limnephilidae family actually crawl or swim along the bottom and emerge from the shuck on land. But the rapid ascent is the most common by far, and the rise forms, as we will see, will usually tell very quickly how the pupa is behaving.

In *The Challenge of the Trout,* Gary LaFontaine reports some interesting observations of caddis pupae underwater. Donning his scuba gear (that's what I call doing your homework!), he has discovered that trout do not necessarily have to take the pupa on its rapid ascent to the surface. Many times Gary observed pupae close to the bottom, drifting freely over the streambed just prior to a hatch. Apparently there is a moment of hesitation between the time the pupa emerges from the case and the actual release of the gases which propel it to the surface. This explains the reason for the effectiveness of the dead drift (fishing weighted flies on a slack line) technique of fishing pupa imitations. And, upon reaching the surface, many pupae must struggle to escape their shuck— another period of vulnerability, which will be discussed more fully in the chapter on fishing adult caddis imitations. But what Gary has reported, in essence, is that trout will feed more readily on pupae (or any other insect form) when they are *easily* available. The pupa's burst from the bottom of the stream does not go unnoticed by the trout, and the fish do take pupae as they rise. But why should the trout hurry when the same meal can be had in a leisurely fashion? If the fish can find pupae uncased on the streambed, they will take them with far less effort and without any hesitation of their own.

Although at the pupa stage the caddis appears to have legs and wings, they are not yet fully formed. The time between leaving the case and emergence is the most dangerous age in the life of a caddis fly: it is extremely vulnerable to trout and other fish, and its only protection is its ability to leave its aquatic environment quickly. That circumstance alone determines a great deal of how the angler must approach fishing this stage.

An emerging pupa shows the short, dark, downward-slanted wings typical of caddis, and good caddis pupa ties.

fishing the pupa

Experience has proven that unless you have a full-fledged hatch at hand—minor as it may be—fishing the pupa is generally unrewarding. This does not mean that you cannot catch trout, but it *is* rare, usually a miscalculation made by smaller trout. However, once a hatch is in progress most feeding will *initially* be on the pupal form of the caddis—an opportunity for some very exciting fishing. (Note: a returning *ovipositing* female of some species is often mistaken for an emerger. This phenomenon will be fully discussed in Chapter 5).

If the water you are fishing is less than three feet deep you can usually see the rise of the trout since the rapidly emerging pupa will hit the surface before the fish can get to it, making for splashy rise forms. However, in deeper water where the pupa has a longer way to travel, it is not uncommon to see many insects coming off, but actually fewer rises. The trout has more time, in this situation, to capture the pupa before it reaches the surface.

I recall an incident on the East Branch of the Delaware River in New York when *Brachycentrus* (locally called the shad fly or grannom) were emerging, but only a few small fish were seen rising. The run of water in which I was working was approximately five feet deep. After no response from a caddis dry fly, I switched to a large stone-fly nymph I knew was also prevalent at the time. Within a few casts I had a fat sixteen-inch-plus brown trout. Though I release most of the fish I catch, now and again I will keep one for dinner; I did so in this case. After killing the fish I examined its stomach contents. I counted approximately seventy-five of the similar stone-fly species that had been taken by the fish not too long before. But surprisingly, the fish had accumulated about two hundred of the emerging caddis pupae upon which he had been currently feeding. This fish had probably taken the pupae when they were three-quarters of their way to the surface. I never saw the fish break water in the form of a rise once. After creeling the trout I switched over to a double pupa arrangement at the end of my leader (see "casts") and took five more respectable fish.

Incidentally, larger fish *especially* do not like to expend more energy than they have to in order to feed. They will usually take the easiest prey available. In reverse, smaller fish will dash hither and thither trying to catch anything in sight. Many times they only succeed in creating a lot of commotion.

twitching rod technique

Because the pupa emerges from the bottom in a unique manner it is important to imitate not only its size, shape and color, but also its *motion.* I have found it important to use flies that have been weighted. If you prefer not to weight your flies, you should at least have some weight on your leader; otherwise, it will be difficult to obtain the rapid ascent from the bottom to the surface which is characteristic of the natural pupa, because your fly will not start on the streambed.

For best control, fish with a relatively short line. The cast should be made up and across stream. *The fly and leader should be allowed to drift downstream with no drag.* If you have to mend your line a number of times to obtain this effect, by all means do so. When the fly gets below you and begins to swing help it along by lifting the rod tip gently in a pulsating movement. Ninety percent of your strikes will occur at this time. This cast effectively imitates the moment of hesitation after the pupa emerges, and its sudden and rapid ascent.

One of the best caddis pupa anglers I know, John LaFonte of Hawthorne, New York, has built quite a reputation on the Housatonic River in Connecticut using this technique. This river has a predominance of grey caddis hatches. During the emergence John would outfish most anglers on the stream by at least five-to-one. He uses a unique woven fly which his friends have teasingly named the LaFonte Wobble-Rite. It sounds more like the name of a bass lure than a trout fly, but a fly it is and a good pupa imitation at that. It has been labeled "Wobble-Rite" because of the wobbling action it imparts as it rises to the surface on the upswing of the cast. (This pattern is illustrated on page 84.)

This is the way John fishes his imitation: he picks a rising fish and strips out just enough line to reach it. If possible, he always tries to keep the distance between himself and the trout under thirty feet. He then makes an upstream cast into the same current lane in which the trout is feeding and manipulates the fly in a dead drift down to the fish, allowing it to sink as deep as possible. When the fly begins to swing in front of the trout John raises the rod, continuously twitching the tip (much in the manner of a puppet on a string) giving a lifelike action to his imitation; *smash!* was the result—most of the time. Don't expect gentle strikes or takes when fishing in this manner. You may, in fact, have some flies broken off if you react too sharply; you must get accustomed to this technique. The LaFonte method is applicable to whatever caddis pupa imitation you use, though the Wobble-Rite has just a bit more wobble.

casts A "cast" here is an arrangement of two or more flies tied to the end of a leader. The additional flies, other than the fly at the tip of a cast, are referred to as "droppers." Casts are not as common today as they once were, during the heyday of the wet fly. It was once the impression of many anglers that wet flies were imitations of drowning or spent wing mayflies, or in some cases, emergers. To an extent they were. We know now, however, that trout take many of the wet fly imitations for caddis pupae. Two of the most popular wet flies ever fished are the Leadwing Coachman and the Gold-Ribbed Hare's Ear. Think of the colors that go into the make-up of these patterns. They are brown, tan, olive, and grey. Those are the predominant shades of the various caddis pupae.

Wet flies were often fished in a cast of three flies, all of them different. One, however, was usually an attractor pattern, such as a Royal Coachman. The attractor was always the first, or highest up, on the leader. This fly, though rarely taken, would get the attention of the trout while one of the two remaining would be close in color and characteristics to the natural upon which they were feeding. If one fly was found to be the producer, the angler then of course concentrated upon that pattern.

Casts for caddis pupae are arranged in much the same manner, except that no attractor pattern is employed. Usually three different types of flies are tied to the leader. A typical arrangement would be a Jorgensen Caddis Pupa, Solomon's Caddis Pupa and a LaFonte Wobble-Rite. The latter would be tied highest up on the leader for though it is not an attractor as such, it certainly would attract more attention than the others.

If the leader is knotted, loop snelled fly above knot and pull tight.

Remember, make dropper leader stiff and less than 4" long.

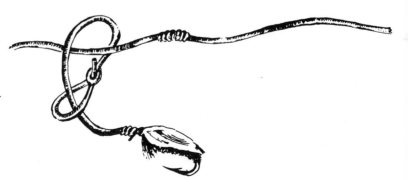

Fishing a cast in this manner will also tell you at what depth the fish are feeding, according to which fly is taken, in addition to telling the preference for a particular pattern. Once a pattern preference has been established, the angler can remove the ineffective patterns and concentrate on a one-fly arrangement using that pattern which has proven successful.

how to arrange a cast of pupa

Many anglers have their own way of splicing tippet material to leaders for making casts, using different knots, and various lengths of tippet material. The important thing to remember is to make your cast for the type of water to be fished and to keep the arrangement so that the flies do not tangle with the leader unnecessarily.

If you are unsuccessful with the particular pupa cast you are using, try changing the sizes of the various flies. Sometimes it will be the very small (or very large) natural on which the fish are feeding. Finding the right formula becomes a matter of experimentation. Again, these are the methods usually preferred when insects are emerging but no trout are seen rising.

Fishing the pupa is also effective during surface feeding, especially at the beginning of a hatch. Capturing and examining some of the insects that come off will eliminate much of the guesswork in the selection of patterns.

If leader is knotless, make loop in leader as shown above.

Attach snelled fly to loop.

salmon fishing cast

If you are a salmon fisherman, or familiar with the ways in which most salmon are fished for, you can also use this salmon fishing technique; it will cover a good deal of water for you and is effective when no fish are seen rising. Very simply, the casts are made downstream and across, at an angle of approximately forty-five degrees. After the fly alights on the surface, it is allowed to sink and swing in an arc until it is directly downstream from where you are standing. The only way this cast differs from a true salmon cast is that a few feet (five to six) of line are mended upstream in loose coils when the fly first hits the water in order to allow the imitation to sink before it begins its swing. If there is no strike, take a step or two downstream and repeat the cast. Again, use a line of no more than thirty feet, if possible. Using this method you can cover an entire stretch of water, and discover where some good trout lies are in the process.

Each time the fly swings in the current the rod tip should be twitched. Make your pupa imitation appear to be alive!

tying the pupa

Three very effective pupa patterns and the instructions for tying them are described in the following pages. This does not mean that there are not other highly effective local patterns tied by anglers throughout the country (see chapter 7). These undiscovered patterns have just never had a chance at popularity simply because no one ever wrote about them. If an off-beat pattern you have been using works for you, by all means use it. Just remember: if you want to increase your success ratio, keep in mind the natural you are imitating, and match it for size and color. Again, if you are trying to match a natural's coloration in a pupa or larva pattern with fur or wool dubbing, be sure you compare the *wet* color of your material.

If you have never tied or used a caddis pupa imitation, these three patterns will give you a variation of tying techniques and a variety of types to use on the stream. All of them may be changed in size and color to match your natural. The emergence tables in Chapter 6 list the adult colorations, which will always approximate the pupa, except that the pupal wing slats should be somewhat darker. The legs and body will remain the same. These patterns use a dark complementary thread to match the slightly darker head of the emerging pupa.

The wings should extend one-half the length of the hook past the bend, and slant downward. If you are tying from the emergence tables, you can use the same hook size as is listed for the adult.

For materials for the various bodies there is a new kind of knitting yarn on the market referred to as "sparkle yarn." It is offered by various manufacturers in different colors under a variety of names (one brand is called Frosted Wintuk); all of them are acrylics containing a silvery flecking which makes them sparkle. Now, when natural caddis pupae emerge, the bubbles emitted from the shuck give off light reflections. The sparkle yarn gives a similar illusion. I have fished it with surprising success and am inclined to believe it may be the most suitable material of all for pupa imitations. My own supply was purchased at a local five and dime store, but I'm certain that it will soon be available from some of the materials supply houses. Sparkle yarn comes in various colors that can be blended in order to obtain a preferred shade. Although the illustrations for the following flies were not tied with this material (I discovered its effectiveness after the photographs were taken) the pattern description lists sparkle yarn as a body substitute.

In his own book, *Challenge of the Trout,* Gary LaFontaine has given it much broader mention and significance for fishing the pupa. In fact he states that pupae should be tied without any wing slats at all since these portions of the pupa are not readily seen by trout. To quote him, "the wing slats are not prominent because they are inside the air-filled pupal sac. The glitter of the air is the 'trigger characteristic' that initially causes the fish to chase the insect." All of us have been aware that pupal imitations as described are effective producers. On the other hand so are many wet flies. The Gold-Ribbed Hare's Ear has no wing slats, and yet many anglers claim it is their most effective caddis pupa. The Hare's Ear also is made from rough textured fur, which in turn traps minute air bubbles.

Two of Gary's flies illustrating his use of the sparkle yarn in the pupa and an emerging pattern have been listed in Chapter 7.

jorgensen fur caddis pupa

hook: Mustad 3906 or 3906B

thread: Dark brown

body: Brown wool or fur dubbing (rabbit, raccoon, or sparkle wool), or the appropriate shade to match the natural

wings: Sections cut from mallard or black duck feathers

thorax: Dark brown or appropriate color dubbing (muskrat, mink, or similar furs)

legs: Guard hairs from same fur used in thorax

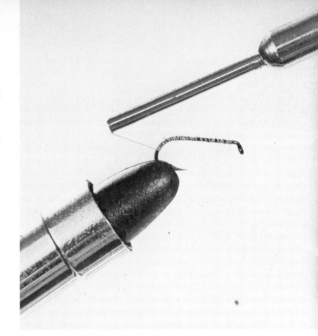

1 This pattern is tied in sizes eight through fourteen. Place an appropriate size hook in your vise and spiral a dark brown fine thread onto the shank commencing just behind the eye and wind to the bend. Make sure the thread is worked well down into the bend.

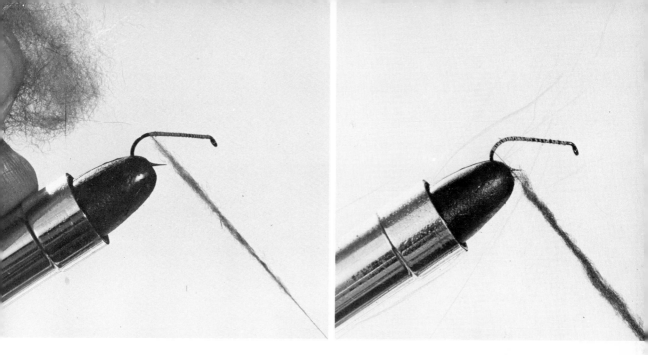

2 From a piece of rabbit, or similar fur, dyed a medium dark brown (or whatever body color you are imitating) pluck a fair portion of underfur. This will be dubbed onto the thread to form the body. If you are familiar with the noodle or rope method of spinning a dubbed body, by all means use that, since it will give you more of a segmented effect. (If you wish to weight your fly, see the Latex Caddis Larva for procedure.)

3 Spin the underfur onto the thread.

4 Wind the dubbed thread forward to a point approximately one-quarter inch before the eye of the hook. The rear portion of the fly should have slightly more fur forming the body than that portion closest to the eye. If, at any time, you need more dubbing to accomplish that effect, by all means add to it. It's good to remember that it is always easier to add more dubbing than to have to remove an overabundance. The body of the fly at this juncture should appear as in the photograph.

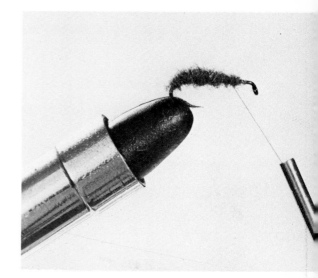

PUPA

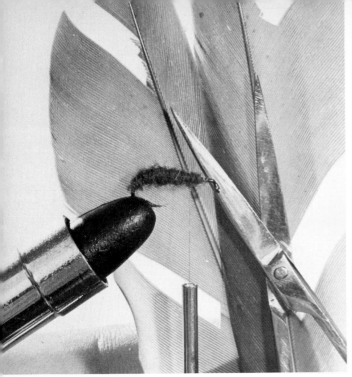

5 For the wings you will need a pair of dark matched mallard or black duck pointer quills. They should be dark.

6 From each of the quills cut a section of fibers not quite an eighth of an inch wide.

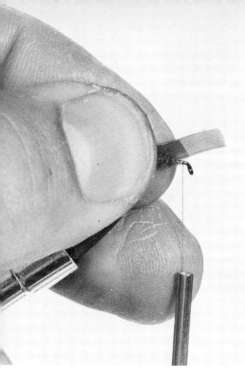

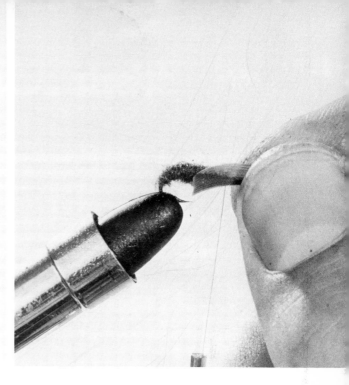

7 Place one against the far side of the hook shank and tie it in. Clip the excess protruding out past the eye of the hook.

8 Place the other section of quill on the near side of the hook shank. Measure each quill so that the tips, which should curve in a downward slant, extend approximately two thirds of the body length toward the bend, and below the body plane.

9 Tie in the near quill and clip the excess. It should appear as in the photo.

If you have any difficulty in positioning the wings properly, or if the wings have a tendency to flare outward away from the shank, the problem may be that you are not tying them directly on and against the fur body. *The quills sections must be tied against the fur body, not the shank of the hook forward of the fur.*

PUPA

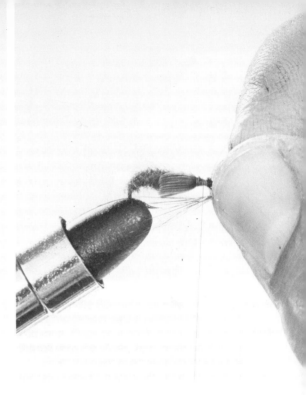

10 The next operation will be the thorax and legs of the fly. For this you will need a piece of the same fur which also contains a goodly number of guard hairs contained in it. Both will be used in conjunction with one another.

11 If you are familiar with the technique you can probably tie in both the underfur and guard hair simultaneously and get the job done in one operation. For most of us, however, it will be easier (and neater) to first grasp some of the guard hairs and tie them in as you would the throat of a wet fly.

12 Once the guard hairs have been secured and the excess butts trimmed, spin some of the dubbing fur onto your thread. A few short guard hairs can be left in with the underfur.

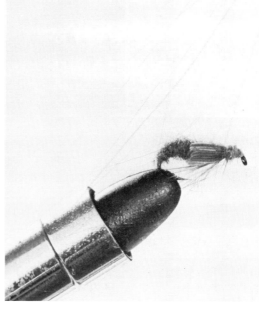

13 Wind the fur and guard hair around the shank of the hook to form the thorax of the fly. Clip most of the fur from the top of the hook shank and stroke the underpart with your fingers so that the fur extrudes down and rearward, thus forming both the thorax and the legs. The completed caddis pupa should appear like this.

The Jorgensen type of caddis pupa may also be tied in other shades, depending on the natural being imitated. Body shades of cream, tan, and olive are very common.

solomon's caddis pupa

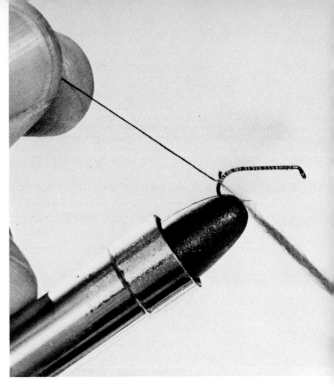

hook: Mustad 9671 or 3906B

thread: Dark

body: Fur, wool, or sparkle wool to match the natural, usually olive green

rib: Dark–colored monocord

wing: Section of quill cut from mallard or black duck wing feathers for sides (for a mottled wing pattern use turkey)

legs and antennae: Brown partridge (or substitute)

head: Peacock or ostrich herl

1   This pattern is tied in sizes twelve through eighteen.

Place a preferred hook in the vise and again spiral the thread onto the shank, terminating well into the bend of the hook. If you wish to weight your fly, now is the time to do it.

Cut a six-inch section of dark monocord and tie it onto the shank well into the bend; leave it idle for now.

From a piece of dubbing fur that has been dyed an olive green (or your preferred pupa coloration) pluck a small amount to be used for spinning onto the thread. Spin the fur onto the thread.

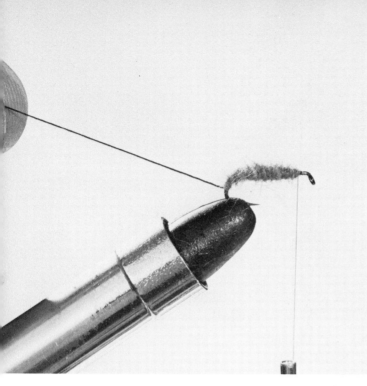

2 Wind your dubbed thread forward to a point one-quarter inch before the eye of the hook. Add more dubbing as needed to form the proper taper of the pattern.

3 Grasp the monocord (brown is good in this tie), which you left at the bend, and wind it in an open spiral in a counter-clockwise motion toward yourself, to the thread. (This will accentuate segmentation.) Tie it down and clip the excess.

PUPA

4 Again select two sections (approximately one-eighth inch wide) of dark mallard or black duck quill and cut them from the respective right and left feathers.

5 Tie one of them in on the far side of the hook shank. Measure the next against the near side of the hook shank. They should extend rearward and in a downward slant covering two-thirds of the body of the fly. Tie in the near wing and secure it. Clip the excess butts.

 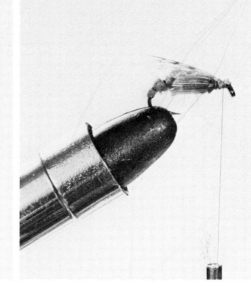

6 Select a brown partridge feather, or if you don't have any of the natural, a mallard side feather that has been dyed to imitate the right shade. (Hen hackle fibers will do, though they lack the barring effect present in the feather).

7 Cut a few fibers (about six or seven) from the partridge feather or hackle and place them on top of the hook shank. Measure them so that the tips extend to the bend of the hook. Tie in the fibers and clip the excess.

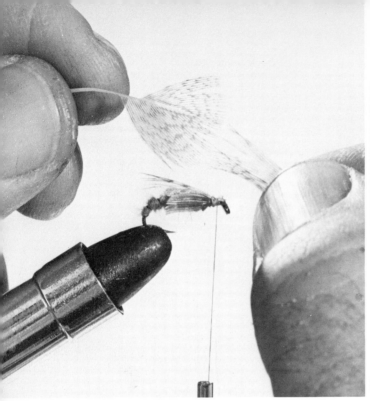

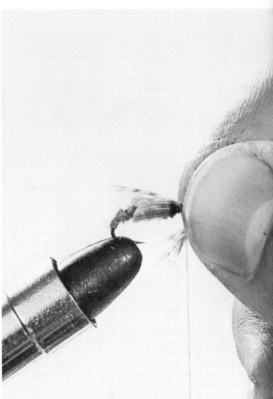

8 Cut another small section of partridge or hackle fibers and roll it into a clump.

9 Measure the second section of partridge fibers along the bottom of the hook shank behind the eye so that the tips extend to the point of the hook. Tie them in as you would the throat of a wet fly. Clip the excess butts.

10 Cut a wide-flued peacock herl fiber from an eyed peacock tail feather and tie it in just behind the eye of the hook.

11 Wind the peacock herl around the shank so that it forms a full and bushy head. A touch of head cement and a whip finish with the thread completes the fly. Solomon's Caddis Pupa should appear as shown.

   Again, this pattern can be tied in various body colors and sizes according to the natural you want to imitate.

la fonte
wobble-rite

hook: Mustad 3906B

thread: Dark

body: Use various combinations of light and dark floss woven to match the natural.

underbody: two strips of fine lead wire

wing (representing antennae): A few partridge or wood duck fibers

side wing (or legs): Partridge hackle

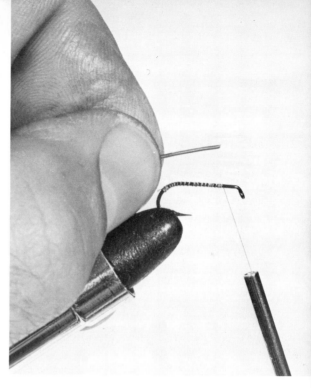

1 This pattern is tied in sizes ten through sixteen. Place the right size hook in your vise. Spiral your thread onto the shank beginning at the bend and winding forward to a point approximately one-eighth of an inch before the eye of the hook.

   Cut a piece of fine lead wire and measure it against the shank of the hook so that it covers the shank from the bend to the thread.

2 Cut another section of lead wire the same length as the first. Tie one piece on each side of the hook shank and lash them down securely with your thread. A liberal application of head cement to the windings will help keep the wire in place.

3 For the woven body you will need two spools of single-strand floss. We are using yellow and olive; the right color combination depends on your natural.

4 Cut a ten-inch section of floss from each spool. Wax each section of floss by running it through a ball of wax a number of times. Lay the ends of both pieces of floss on top of the hook shank and tie them in with the thread.

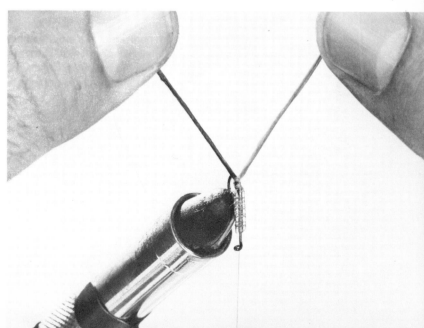

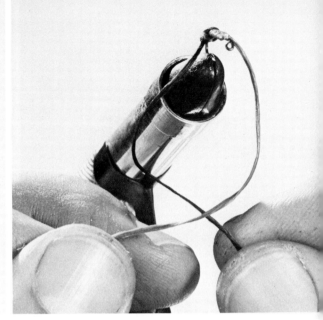

5 Bring your thread forward to the eye of the hook, half hitch or whip finish, and cut the thread from the hook. It will only be in the way for the weaving operation. The olive floss should be coming off the hook at the bend on the left side, and the yellow on the right.

6 Bring both strands of floss under the shank of the hook by holding one strand in each hand. The yellow floss which will be the underbody, should cross over the olive in front of it.

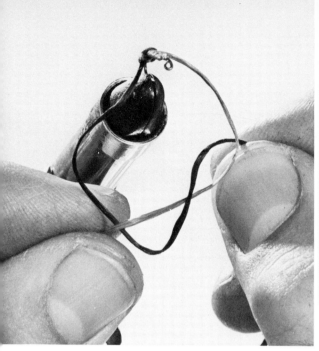

7 Bring the olive floss through the yellow floss in the manner of a simple overhand knot.

8 Slip the open loop, or overhand knot which has been left open, over the eye of the hook keeping the olive (dark) floss on top and the yellow (light) on the bottom.

9 Bring the loop all the way to the rear of the bend and then bring it tight. Keep repeating this process and gradually build the body of the fly. Always remember to keep the dark floss on top and the light floss on bottom. Keep the yellow (underside) crossing in front of the olive (top).

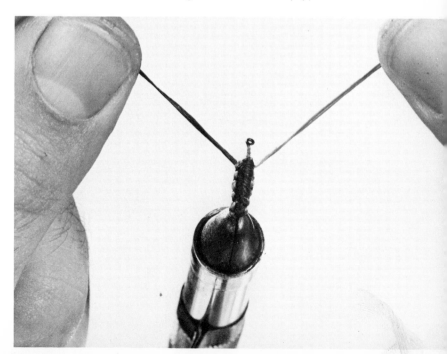

PUPA

10 Build your woven body to a point one-eighth inch before the eye of the hook, or until the underbody of lead wire has been completely covered.

Spiral the bobbined thread onto the hook shank again.

Bring the floss slightly forward and bind it down with the thread; clip the excess.

11 Cut a section of wood duck or partridge fibers from their respective feathers.

pupa: points to remember

1 Splashy rise forms are generally indicative of trout feeding on the pupa just prior to emergence.

2 Pupae move to the surface rapidly; your imitation should be fished in a like manner. Twitching and raising the rod tip as the leader swings in the current imparts such an action.

3 Check the depth of the water you are fishing; if it is more than three feet deep but no rise forms are evident and there are adult emerged caddis on

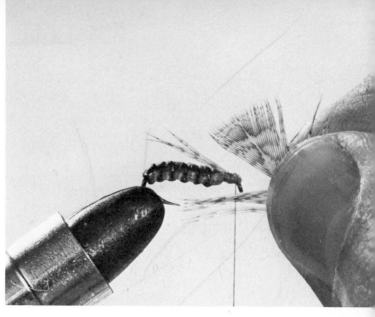

12 Tie in a half a dozen fibers to form the wing (antennae) on top of the hook shank. The tips of the fibers should extend to the bend.

13 Cut another section of fibers, using about twice as many as for the wing, brown speckled partridge or grouse feather, and measure it as you would the throat of a wet fly against the hook. The tips should extend to the bend of the hook.

14 Taper a head with your thread, whip finish and add a touch of head cement. The completed La Fonte Pupa should appear like this.

the wing, you can be almost certain that trout are getting to the pupae before they have a chance to break through the surface. If the water depth is less than three feet you will generally see the splashy rise form.

4 Fish a length of line and leader you can control, preferably under thirty feet.

5 If you cannot determine exactly what kind of pupa trout are feeding on, or where they are feeding, use a cast arrangement of several imitations and cover the water as you would in salmon fishing.

the adult caddis

imitation and presentation

Upon hatching from the pupal shuck, the adult caddis wastes no time in getting airborne. A strong flier, it is usually seen performing an erratic flight pattern over the water; it would seem that it is constantly trying to dodge something. Newly emerged adults quickly fly to land, where they rest and dry their wings, which slant back over the body when they are not in flight, like a long tent. Thus in both ways caddis are easily distinguished from the gentle-flying, upright-winged mayfly.

It's difficult for the fly fisherman to judge the size of a flying caddis since its four full-size fluttering wings give the illusion of a much larger appearance than the actual body size. Here it differs again from the mayflies, whose body length is about the same size as its main two wings. Caddis have a much shorter body-to-wing ratio: generally, the body of an adult caddis will be only two-thirds the size of the wings. Caddis do not have the elongated tails that most mayflies show.

Since most caddis also appear lighter in color in flight than they really are, it's a good idea to catch a natural to really match the hatch. As we shall see, fish can be extremely selective during a caddis hatch. But caddis are by far the most difficult stream insect to catch for identification. The crazy flight pattern makes them as elusive for the angler as they are for the trout. Should you decide to do a little collecting and examining on your own, a small four-by-six-inch net fixed to a handle is essential. While it is not difficult to catch mayflies, caddis will prove a challenge.

Again unlike most mayfly species which have a life span of only one to three days after emergence, the adult caddis may live up to eight weeks, and most have a life span of at least two. During most of this time they will not be readily available to the trout. But besides their vulnerability to fish at emergence and—for females—during ovipositing, the adult caddis return to the stream periodically for liquid nourishment, the only kind their adult mouth parts can take in. Trout are aware of this sporadic presence and thus may be taken by proper presentation of a dry fly during seemingly slack periods when there is no real hatch activity.

Unlike the transformation of larva to pupa, during which the body color often changes considerably, the newly emerged adult will have the same coloration as the advanced pupal stage (which makes tying the pupa a little

Caddis have a much shorter body-to-wing ratio than mayflies.

A fully emerged adult caddis, *Psilotreta labida,* an Odontoceridae family member; caddis adults will often mate two or three times in their ten to twenty days as winged insects.

easier). However, once the insect has been exposed to the atmosphere for a period of time, its coloration may begin to take on a darker shade. Then later, the female may lose some of her abdominal coloration and even diminish in size after she has deposited her eggs.

The caddis will spend most of its adult days in dark places along the stream, among rocks and foliage. The insects will fly about quite freely at dusk, often in large masses. These flights may take an upstream direction, or they may be observed just skimming back and forth above the surface. This type of activity usually excites the angler more than the trout. It is an indication of mating time; it should not be confused with ovipositing, which generally takes place after dark. The ovipositing characteristics of the caddis can be especially important to the fly fisherman, and this stage has a chapter of its own in this book, but it's worthy of note here that caddis may mate two or three times before they expire. The actual mating will take place over land, where the caddis will also die.

Much of this emergence activity goes unseen by the angler since it takes place at dusk or in total darkness. Fortunately, enough hatches do occur during the daylight hours to make fishing this insect as a dry fly quite rewarding. It is one of the most exciting artificials to fish because of the sharp, splashy rise that results from its usually rapid emergence pattern.

Masses of *Brachycentrus*
adults in an upstream
mating flight on the Beaverkill
in New York State.
The slow camera speed
accentuates the blur of
motion which characterizes
adult caddis flight, and
often makes the insect seem
larger than it actually is.

fishing the caddis adult

The most important thing to do when fishing with a caddis imitation is to *present your fly properly.* This, of course, holds true for all types of fly fishing. However, for fishing the caddis dry proper presentation requires the use of some techniques not used in conventional dry fly fishing.

But what *is* proper presentation?

It is the correct and controlled manipulation of rod, line, and leader which allows the fly to float on the surface of the water in a manner that will stimulate the interest of the trout. In short, the ability to cast well in various ways.

However developed your ability, you will do well if you remember the limit of your control of the line. In other words, if you can handle and control only thirty feet of line, do not fish forty feet of it. You will succeed only in putting down more fish than you will raise. A fly dragging across the surface in an unnatural manner is one of the major reasons trout refuse a dry fly.

Casting technique is an overworked angling subject, and yet, the fact remains that if you know how to handle a rod well and make the proper cast in a given situation, you have a much better chance at success than the average angler, even if his imitation is a more perfect match than your own.

Many excellent books have been written on casting technique alone; and nearly every general fly fishing book contains the basics of good casting: timing and direction of the rod tip. Still, many anglers and especially novices find it hard to learn casting from books and there probably is no real substitute for firsthand instruction. Enlist the help of a good caster—friend or streamside acquaintance—or take a course in fly casting from one of the many commercial clinics to achieve the basic level of control. After this kind of instruction the casting techniques presented in books become even more valuable, and the effect of the two together is greater than the sum of their parts. And once you have learned the basics, keep your casting arm in tune by practicing what you have learned—especially during the off-season.

Our main concern in *The Caddis and the Angler* is to offer a few types of casts that are necessary in order to fish the caddis dry fly successfully. These casts, some old, some new, all assume a certain amount of control. They may not be difficult in themselves, but imitating the behavior of the natural caddis adult with a dry fly often calls for precision at some point in the cast. A good grasp of casting basics and a fair degree of line control skill is essential.

the flutter

The flutter is not really a cast at all, but a type of action imparted to the fly. As implied by its name, the flutter gives the fly an erratic skipping or fluttering motion after it alights on the water. This action is not brought about by the cast as such, which can be an ordinary and conventional everyday type of dry fly cast, but by the fingers and wrist of the casting arm *after* the fly has landed. Accomplished correctly, the fly should bounce, skip, and flutter on the stream surface much in the manner of a skater or spider type of fly.

Although the action of the flutter cast has been utilized by many anglers over the years (often referred to as "skating"), Leonard Wright, in his book *Fishing the Dry Fly as a Living Insect,* was the first to illustrate it, and bring it to the attention of anglers everywhere. He referred to this action as "the sudden inch".

The cast is always made downstream, or quartering downstream, and the fluttering motion is made to bring the imitation upstream in the same manner in which an emerging natural would behave.

It is important that a good cast be made, giving you direct control and contact with your fly. Uncontrolled slack line will not allow you to perform the function properly.

For example: You are presenting a caddis imitation to a rising trout thirty feet down and across stream from your position. The cast should be made so that your imitation lands approximately two or three feet above, and in the feeding lane of the fish. The rod tip is held fairly high and the tip is twitched in a staccato movement, while the rod arm gently twitches the fly into the line of drift in which the trout is rising. This motion should not be overdone; a "flutter" is not a yank. The fly should move only an inch or so at a time. There should be a pause during which the fly is allowed to float naturally. If no results are obtained on the swing, another twitch or two may bring a strike.

Trout are familiar with caddis. There is an instinctive "don't-let-it-get-away" reaction to an insect skipping and bouncing along the surface, and, if your imitation is performing properly the trout won't let it get away.

upstream hold or extension cast

The trouble with most streams, for the average fly fisher, is that the flow of water is never of a uniform speed from one side to the other. Usually the center of a stream flows faster than the sides, but there are in many stream situations you may face as many as four or five variations of current speed at a given point.

The problem this presents to the angler of course, is the ability to obtain a drag-free float. The conventional straight-across stream cast won't do it since the center current will whisk the heavier line downstream faster than the leader and the fly. Therefore, to offset and counterbalance the faster stream flow most anglers will, after the cast has been made and the line is on the water, mend their line so that it falls upstream of the current, attempting to give the leader and fly a headstart and assuring a certain number of feet of natural floatation. Unfortunately, it usually does not work. Mending the line causes loss of control and contact with the fly, and the fly is pulled away from the line of drift, generally creating an unnatural movement.

Remember, in order to properly control the fly while imparting a fluttering action, your fly should be cast down, or down and across, to the fish. By fishing downstream to the fish, the fly will get to him before the leader or the line.

ADULT CADDIS

The best method I know which assures the most line control is a cast called the "upstream hold", or "extension cast." This technique will allow the fly to land directly on its intended target without having to overcast or make unnecessary mends or movements after the fly is on the water. Allowance for adjustment of the line is taken into consideration *before the fly is cast.*

Here's how it works. There is a trout rising twenty-five feet down and across-stream from your position. The middle and far currents are moving slower than the water you are standing in. You must cast two to three feet above the lie of the fish and get a drag-free float to approach him naturally. You will need about five or six feet of shooting line to compensate for the extension or hold you are going to make to keep the main part of the line upstream and give your imitation a head start.

False cast your line so that your fly is directly over the target, to measure the amount of line to work in the air. Now strip five to six feet of *extra* line from your reel and keep it loosely beside you.

Make the cast as you normally would, by pointing the tip of your rod to the target area, which will give your line and fly direction.

As the normal forward cast is unfolding, pull the extra five or six feet of loose line (which you held) to slide through the guides as you bring the entire rod, butt leading, then tip, *upstream* (to the right or the left, depending on which side of the stream you are fishing) as far as you can. This should be accomplished in a smooth and fluid motion; there is no pause throughout the execution of this cast. The five or six feet of loose line which has been allowed to slip through the guides during the maneuver of bringing the rod upstream, acts as a slide. If you did not allow the loose five or six feet of line to pass through the guides as you moved the rod upstream, you would pull the fly away from its target.

As your fly and line move downstream in a dead drift, follow the movement with the rod at the same speed as the current. This will enable your fly to drift drag free, and at the same time allow you to have direct contact without any slack line, so that you can strike in the event of a rise.

A right-hand caster making the extension cast with the current moving from left to right.

The forecast.

The line unfolds; the angler begins to move the rod upstream.

Drawing the line through the guides to bring the rod into the farthest upstream position; the angler can

now maintain contact with the fly as he follows it downstream.

When the current is moving from left to right, a right hander will have to make a somewhat more limited sweep across his body to pass the extra line through the guides. As you can see from the last two photos in the sequence, this cast is easier (for a right hander) when you are casting across current moving from right to left. In this situation, you need only extend your casting arm away from the body.

The extension cast can also be executed directly downstream, if you see a riser in that position. Make your cast as before, and draw the extra line through the rod directly towards your body, as in the last photograph. Then you can follow the drag free float of your fly downstream over the target. The beauty of the extension cast is that, in all three instances, it allows you to get maximum drag advantage from the rod and line position, without losing direct contact with the fly, even momentarily.

mend-and-skitter cast

Although I do not recommend any type of mending cast, there are times when a situation presents itself so that it is impractical to use the simple and effective upstream hold cast.

The mend and skitter cast might at times be re-named a "desperation cast," since it turns the trick when all else has failed. It has, however, a very functional purpose, in addition: in a crude manner it will give you a drag-free float. It is relatively simple.

The fly and leader are cast beyond the target area. As soon as both line and leader have made contact with the surface of the water, the line is immediately mended toward the upstream side of the current. As you make the mend, which should be a substantial—almost violent—one, you will notice that this action on your part causes the fly to skitter across the surface towards you. If you have estimated the distance correctly the fly should begin a natural drift just above the desired target area, whether to a rise form, or a desired lie. The fly will behave like panicked caddis, which is exactly what you want. The upstream mend (which resulted in the skittering motion) will allow you a fair distance of natural drift.

The finish position for the same angler casting across water moving from right to left.

The finish position for the downstream extension cast. Here again, the angler is now able to present his fly without drag and without losing contact.

tension cast

While the downstream, or down and across stream, cast is the most effective for fishing the adult caddis imitation, there are times when an upstream approach is necessary.

The tension cast is a particular favorite of mine when I am fishing small streams that have an overabundant backdrop and hanging tree limbs. It is also simple to execute. The only difficulty arises in putting the fly exactly where you want it; it requires practice to learn to control the tip of the rod with pinpoint precision so that the fly lands on the target.

Perhaps the best way to describe the tension cast is to cite an example. You are fishing a small stream which is heavily bordered with bushes and there is a fair amount of overhang behind you preventing any backcast. There is a trout rising just below an overhanging tree limb about a foot or so above the surface of the water. You must get the fly above the lie of the trout.

1. Mentally measure the distance from where you are standing and where the trout is rising. Allow the fly line to float downstream behind you. The current will straighten it out. When it has completely straightened and is so taut that you can feel the current tugging against it in your hand you are ready for your cast.

2. Extend your arm rearward as far as possible and bring your rod forward in a side-arm cast. You will feel the pressure of the water loading the rod for you. As you make your power thrust, the line will leap from the surface and skim along about eight inches above it.

3. As you complete the cast point the tip of the rod at the spot you wish your fly to land. Do not raise the tip of the rod higher than you want your fly to travel *at any time*. If you miss your target, it will be because you have not controlled the tip precisely. (Remember: all rod tips vibrate, and you must take this into consideration. The line will go wherever the rod tip has pointed to *last*.)

types of water

Most anglers are familiar with the various classic types of water found in most streams, namely: the pools, riffles, pockets, and glides. Even non-anglers can grasp the meaning of these very descriptive names.

With regard to caddis imitations, these types of water should dictate to you the type of fly you are going to use, and the manner in which you will fish the particular stretch. In essence, it is simply a common sense approach.

For instance, in fast or riffly water you will want a fly that floats higher. This involves tying your favored pattern in a material which is water resistant, or slightly overhackling the fly. It may not have the same definition you would normally give the pattern, but that won't make much difference to the trout since the fly itself will be distorted, due to the nature of the water through which the fish sees it. It will serve the purpose much better, moreover, than a conventional tie. And, of course, the reverse also holds true: a heavily hackled hairwing fly can

be tied sparsely, or a similar pattern which is more realistic in silhouette should be used when fishing in quiet water. Under certain conditions a standard pattern can be reduced on the spot for selective pool trout simply by cutting a "V" into the bottom of the hackle collar. You will find that in slick glide water the trout will be *ultra*-selective regarding size and color.

For the most part, caddis activity takes place in the faster moving, or riffle type of water, though many times the actual feeding will take place in those areas just below. And, there are some species that prefer quiet waters (see Appendices). But for the most part, you will have more types of caddis—and much more response—if you concentrate on the portions of the stream in which there is a good flow of water.

rise forms

Since we will be referring to the manner in which trout will behave under certain conditions, a brief explanation of the types of rise forms is in order. For caddis, we are concerned with just two of them.

In relation to the caddis, the most common rise form is that of the sharp and splashy variety, which is indicative of a fish chasing a rapidly-emerging pupa, or in pursuit of an adult caddis which has returned to the water for liquid nourishment and is fluttering about close to the surface. In both instances the trout simply cannot take their time. If they are going to succeed at all in obtaining food

ADULT CADDIS

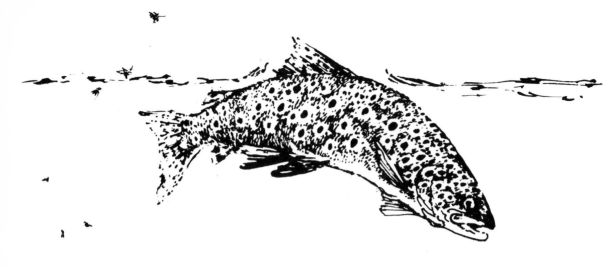

in the form of emerging pupa or fluttering caddis, they have to be quick. The momentum of their charge often carries them completely out of the water. They seem to throw caution to the wind. They create a disturbance. They are noisy.

The less common rise in relation to the caddis is the classic head-and-tail rise, in which first the trout's head, then the tail are clearly visible near the surface. This rise form is generally associated with the mayfly, since this insect is relatively slow in emerging and becoming airborne.

It takes place during a heavy, flush hatch of caddis flies during which many insects are present, especially when there are cripples among the hatch. The occurrence of crippled emergers is common to all aquatic insects. Some cannot extricate themselves from their pupal or nymphal shucks; some emerge not fully formed. The handicap thus prevents them from taking off from the surface, the fish seem to sense their struggles and it is at this time that trout can take their time feeding on caddis, especially larger trout.

On occasion even healthy caddis also come up from the bottom of a stream in a manner similar to many mayfly nymphs, and will drift in the surface film for some distance while emerging from their pupal shuck. These insects also are easy prey for the fish, resulting in a gentle, unhurried head-and-tail rise form. This is the exception for caddis hatches, but the rise form itself, and a few captured emergers, should alert you to the situation. In these instances the conventional dry fly techniques will also serve your purpose in imitating the actions of the natural.

selectivity of trout during caddis hatches

Before getting into more unusual techniques, a brief mention here of trout's peculiar selectivity is in order. I sincerely hope you don't run into ultra-selective fish astream, but if you do, rest assured your technique or knowledge is not necessarily at fault. You may be doing everything correctly—the right imitation, proper presentation—and still you receive refusal after refusal. The problem is again one of observation.

Perhaps an example is in order. The place was Willowemoc Creek in New York. The time was late May. A lovely time, especially during a certain afternoon surrounded by a number of rising trout and the air filled with caddis flies. A time of anticipation and expected results. Expected because I had identified the particular caddis that was emerging.

I tied a proper imitation to my tippet. I cast to the first rising fish and took him. One fish does not a day make, but this certainly was going to be like shooting them in a barrel . . . . or so I thought. I cast to the second fish and took him on the third cast. I had it made. Until the next three rising trout ignored my offering completely. Something was wrong. I again examined one of the emerging caddis . . . aha! It was another species. I quickly changed flies. One of the three trout took the imitation but the other two refused it. I looked around some more and carefully observed the activity. There were three distinct species hatching at the same time: a multiple hatch!

The one thing I did learn that day was that trout are not only selective as a group, they are also individually selective. In other words, while the trout behind the boulder may enjoy *Brachycentrus,* his twin brother two feet downstream may prefer *Hydropsyche.* It is as if they are handed a menu and, as you and I would in a restaurant, simply choose whatever strikes their fancy at the time. Suffice it to say, I had to work for every fish I took that day. I had to find out each fish's *individual* preference.

Trout can be as selective towards caddis flies as they are with mayflies; this selectivity varies with the type of water and the density of the hatch. The fish seem to be most particular in slick, glassy water when there are only a moderate number of flies emerging over an extended period (one to two hours). During a heavy flush hatch or in the riffles of streams, trout are generally easier to take, even if your imitation is not exact. In the case of riffly water the non-selectivity is due to a reduction in the fish's clarity of vision and the fact that it has a limited time to examine the fly in the faster water.

I recall well three caddis hatches I fished during late May and early June of 1973 (all of which later proved to be *Hydropsyche*). There was only a slight variation among the insects. The bodies were a size #18 and had a greyish-olive color. The wings and legs, however, differed, ranging from a tan to a dark rusty dun grey. Having the exact imitation made the difference between taking two, or three, to the dozen I was able to hook on the glassy run of water I was fishing. The trout would swim circles around any fly which did not have the proper wing and leg color credentials, though now and again, one would take it cautiously. On the other hand, when the imitation was correct the take was very deliberate.

During such times, it is the observant angler who is rewarded. I use various detective methods to find out exactly which hatch to match. A small nymph net with a white band of cloth sewn down the middle of the mesh enables me to seine the water itself for pupae and smaller caddis. A small aquarium net (easy to find at pet stores) makes it much easier to capture flying adults. If I intend to keep a fish or two, I will examine their stomach contents by spilling them on a small white plastic tray. Many anglers also use a stomach pump but I strongly advise that this device be used with caution, since it can injure a fish if used improperly. All these items fit neatly into the average vest.

unusual techniques

We have covered some of the general techniques of fishing an adult caddis dry fly, both to rising and non-rising fish. Sometimes, however, standard procedures are not the answer.

A good case in point involves a method used by a very close friend, Ernie Maltz, of Englewood, New Jersey. Ernie has been fishing caddis fly imitations on Catskill waters for over thirty years. He informed me that if a fish is seen rising regularly, you should be ready to drop your fly right on his nose. Even though the trout has just taken the natural, he will usually turn sharply for your artificial if your cast has been presented properly and accurately. The trout's instinctive thought may be, "Ah, there's another one."

The true value of this method is for the times when the trout is chasing the emerging pupa and misses it, which happens quite often. This situation is signalled by a sharp splashy rise followed by an erratic-flying caddis which has emerged from the same spot. At these times the trout will momentarily linger just under the surface looking for the prey that has eluded him. The angler's immediate response in putting the fly right on the nose of the trout will result in taking the fish every time.

Persistence is the key word. I have observed Ernie in the stream making dozens of false casts at a time. During my first observation I was wondering if he was trying to hook a swallow! What he was actually doing was keeping his line, leader and fly in the air . . . at the ready, searching for a target. When the rise did come Ernie's imitation would land in the circle of the rise form within a second, resulting in another hooked fish. When using this method, Ernie advises, it is best to use a relatively short line. This is the most effective way to fish a dry fly when trout are chasing the emerging pupa.

Leonard Wright advocates fishing the caddis dry fly across and downstream, depending on the type of water. This is especially effective when the trout are taking in the classic head-and-tail rise. The approach from above and off to the side allows for a more natural float for imitating caddis behavior with Wright's "sudden inch" twitches. Again, the fly should be presented properly and directly into the line of drift of the feeding trout. I have taken a number of good fish using this method, which again proves that large trout will not move any faster than they have to.

Fishing with a Delta Wing Caddis imitation (see page 134 for tying instructions) at certain times and under certain conditions, will result in more hooked fish than if you were to use the conventional downwing patterns. This imitation is designed to simulate an adult caddis that cannot quite extricate itself from its pupal overcoat, or a partially crippled insect of that species. The wings on this pattern behave in a very life-like manner when twitched, giving the impression of a struggling insect. However, most of my success with the Delta Wing has been fishing it on a dead drift, both upstream and on a slack line downstream.

My first experience with the Delta Wing was in late May of 1972. I was fishing below Painter's Bend on New York's Beaverkill and had noticed sporadic hatching of an amber caddis in about a size sixteen. I was able to pick up a few fish on various patterns, but none of them took fish with any regularity. Rummaging through my fly boxes, I found an experimental Delta Wing tied the season before. With amber body, light rusty dun wings and brown hackle, it had the dominant features of the natural. In the next hour I took trout consistently until the wings came off the fly. Immediately after that I could only raise the occasional fish, and no one pattern took fish consistently.

My next encounter with the Delta Wing came a week later on the Willowemoc, which is in the same water system as the Beaverkill. Jay Herbert, a Texas hunting and fishing companion, was in New York on business and wanted to spend a few days on Catskill waters. Prior to our trip I had tied a half-dozen Delta Wings to match a caddis hatch I knew was due. We were fishing a diminishing grey fox *(Stenonema fuscum)* hatch when the caddis appeared. At first I switched to a Hairwing but failed to get the response I expected. The fish were working well, showing the classic head-and-tail rise. I spotted a very good fish feeding against the far bank but could arouse no interest in the Hairwing. Remembering my experience of the previous week, I tied on a Delta. Before casting to the trout on the far bank, I tested the pattern on two smaller fish rising below me. Each one took immediately.

Fully confident now, I went after the big fish. The first two casts were ignored, but as the third drifted over the lie (the best of my presentations) the trout sucked in the fly and was hooked; it was a beautiful brown fully eighteen inches in length. Jay took one look and demanded one of the flies I was using; "whatever it was." In the next hour we picked up ten trout, half of them twelve to fifteen inches. I was convinced I had a good thing in my strange little bug.

Throughout the season I enjoyed considerable success with the Delta Wing during the majority of caddis hatches. Inevitably, however, there were times when it failed to produce at all. Experience leads me to believe the Delta's success is related to the manner in which some species of caddis emerge from the pupal shuck.

There was some other experience to consider.

During a flush hatch the Delta Wing could be surprisingly effective. If you held a large net in the water you would be astonished at the numbers of partially emerged and crippled insects that remain in the surface film even though large numbers of flies are escaping successfully. As we have seen, these crippled emergers become easy prey, especially preferred by quality-size trout. Of course in a larger hatch there will be proportionately more of these kinds of insects available, and the Delta may well imitate these "cripples" to selective trout.

The Delta Wing also works well during a slack hatching period. Besides resembling a crippled or dead caddis, it is impressionistic of various terrestrials and some spent-wing mayflies. You will find the vast majority of rises to this pattern to be gentle but deliberate, as opposed to the frantic rise trout must often make to take the emerging pupa.

microcaddis In the early days of fly fishing nearly all of the patterns which were tied were dressed on hook sizes twelve or larger. A size fourteen was considered *small.* As we progressed and learned more and more, especially about the mayfly, we found, through our own experience and that of others, that trout do feed upon and prefer the smaller mayflies when no other food is available. In addition, they also enjoy tiny terrestrials as a change of diet during the summer months when these insects are prevalent.

For a while we seemed to be going through the same cycle with caddis flies. The all-time favorite, the Henryville Special, is usually not tied much smaller than a size sixteen, except by a few knowledgeable anglers.

But in fact, many natural-hatching caddis flies are no larger than 5 to 7 millimeters, and they hatch in abundance. Remember, the size of your imitation is the first and most important criterion when you are imitating a hatch. But also important is the fact that the shape or silhouette of the caddis differs from that of a mayfly. Don't think because you have a few midges in your vest that you will fool every trout into taking the offering for a microcaddis imitation.

I recall an instance one mid-August morning when trout were rising but there was no apparent insect life on the water. I proceeded to fish a size twenty-two mayfly imitation which had proven successful during other seasons at the same time of year. My only reward was one ten-inch fish. And yet, better fish were still rising within casting distance. I experimented with terrestrials with little better luck. I scanned the surface again and finally observed two small ginger-colored caddis flies emerging; very shortly thereafter I saw three more. Still, these sporadic emergences would not indicate a hatch . . . or would they? With nothing to lose I decided to try a pattern that imitated them in *size* and *color.* Within the next hour I took and released eight good fish.

Point: When there is a scarcity of food even a small hatch can stimulate trout feeding. The key to trout selectivity, again, is observation and experimentation.

tying the adult caddis

Beyond basics, most fly-tying is just plain common sense. In the case of dry fly caddis imitations the only thing you have to remember is to use materials that will float well, while meeting the requirements of size, silhouette, and shade in a life-like manner.

The instructions for the adult caddis patterns are for flies that have proven themselves. The materials and hooks can be substituted, or changed as long as they meet the four essential requirements. What you will find here is a foundation; any improvements you make can only add to their effectiveness.

You may find it useful to consult the emergence tables in Chapter 6 for color combinations and hook size to match those naturals which may inhabit streams in your region. These charts can't take the place of firsthand knowledge of your waters, but by using the wing, body, and leg color information you find there, you can go astream with as much assurance as the caddis allows. Caddis hatches are hard to predict, but this information will get you started tying (or buying) the most likely imitations, which is of course a pleasure in itself for the ardent fly tyer.

Use a thread to complement the overall color pattern of your fly. For tying from specimens, you can use the hook to insect size correlation chart at the end of this chapter. Remember a caddis looks larger in flight than it actually is, and size is a primary consideration in matching any caddis hatch.

The following patterns illustrate a variety of styles. Let's start with one of the old-time yet still extremely effective favorites, the Henryville Special.

henryville special

hook: Mustad 94840, 94833 or similar

thread: Grey, fine

body: Floss to match body of natural, usually olive

rib: Grizzly hackle palmered

underwing: Wood duck flank fibers

wing: Sections of quill from mallard pointer feathers

hackle: Brown, or quality dry fly hackle to match legs of natural

1 The Henryville Special is usually tied in sizes twelve through eighteen; size fourteen is the most common. Place your hook in the vise and spiral your thread onto the shank, terminating at the bend.

From the neck of a barred Plymouth Rock (grizzly) select a hackle with a fiber length as wide as the hook gap. Tie it in at the bend of the hook.

2 Cut a single-strand piece of olive floss six inches long and also tie it in on top of the hook shank at the bend just in front of the grizzly hackle.

3 Bring the thread forward toward the eye of the hook for approximately two-thirds of the length of the hook shank.

Wind the floss to the thread, bind it down and clip the excess.

ADULT CADDIS

4 Grasp your grizzly feather with a pair of hackle pliers and wind it in an open spiral through the floss to the thread. Tie it down with the thread and clip the excess tip.

5 Cut a V-shaped notch in the hackle protruding above the shank of the hook (about 60 degrees) to allow room for the underwing to be tied in next.

6 Select a finely marked lemon wood duck flank feather, or substitute, and cut a small section of fibers.

7 Measure the section of wood duck fibers along the top of the hook shank so that the tips extend almost one-quarter inch past the bend.
　　Tie the wood duck fibers to the shank of the hook as a single unit at the point where the turns of grizzly hackle end. Trim the excess butts.

ADULT CADDIS

8 For the wing you will need one section of quill from each feather of a matched pair of mallard duck flight feathers. The section cut from the feather should be approximately one-eighth inch wide. Incidentally, the width of the quill sections get narrower as the size of the fly decreases, since in fact, all proportions change.

9 Measure the matched quill sections so that the tips align evenly with one another. Now place them on top of the hook shank and measure them to the length of the hook shank extension. They should curve directly opposite to the bend of the hook.

The Henryville Special is also tied in other body color combinations to match brown and rust-colored naturals. Another very popular version uses fluorescent green floss for the body.

10 Tie them to the shank of the hook by passing the thread between your fingers and the shank and over the duck quill sections; perform this operation two or three times.

Clip the excess butts.

11 From the neck of a medium-brown rooster neck pluck a feather on which the length of the hackle fibers will be slightly longer than the grizzly hackle previously tied in. Trim the base of the hackle feather and tie it in by the butt.

Grasp the tip of the hackle feather with your hackle pliers and begin winding the hackle as a collar around the hook shank.

12 When the hackle collar has been completely formed, trim the excess tip, whip-finish the fly, and add a touch of head cement to the windings. Your completed Henryville Special should appear as in this photo.

ADULT CADDIS

solomon's hairwing

hook: 94840, or equivalent

thread: Fine, waxed; grey is good for most patterns

body: Dubbing fur (mink, beaver) to match body of your natural, usually light olive

wing: Deer hair to match coloration of natural, usually tan

hackle: Brown, or quality dry fly hackle shaded to match legs of natural

1 Solomon's Hairwing is tied in sizes eight through twenty-two. Place a preferred size hook in your vise and spiral some fine thread onto the shank beginning just behind the eye and terminating at the bend.

From some mink fur or similar dubbing that has been dyed the proper shade (light olive is common), pinch a fair amount of the underfur as shown, and spin it onto your thread which is being held taut by the bobbin.

Wind the thread forward toward the eye of the hook forming the body. The body of the caddis should have a reverse taper. In other words it is heavier near the bend than at the eye. Terminate the body approximately three-sixteenths of an inch before the eye.

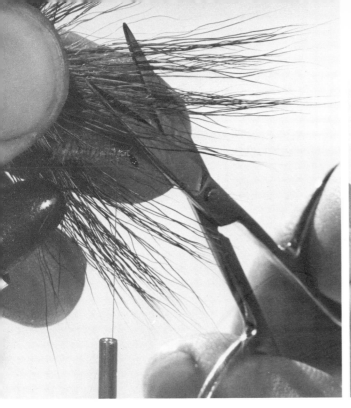

2 From a piece of deer body hair cut a section of fibers (approximately twenty). If you can find a good piece of leg or head hair you will have an easier job of it. The hairs from this part of the animal do not flare as readily as the body hair, but short body hair will do.

3 Manipulate the fibers so that the tips are evenly aligned, or place them in an empty lipstick container or an empty cartridge case (.357 magnum case is perfect).

4 Tap the case on a hard surface. This will bring the tips into alignment.

ADULT CADDIS

5 When the tips have been aligned remove them from the case carefully and measure them along the top of the hook shank. They should extend half the shank length past the bend. Remember, this ratio increases with the size caddis you want to imitate, so refer to the insect to hook size chart in this chapter for larger flies.

6 Transfer the clump of fibers to your left hand, being sure to grasp the hair firmly at the exact point at which they are to be tied to the hook right behind the eye.

7 Clip the excess butts right against your fingers, taking care not to lose your finger position on the hair in the process.

8 Tie in the deer hair by taking two or three turns up through your fingers and around the fibers; then roll the tips of your fingers away from the shank slightly, and tie the hair down with two or three more tight turns of the thread; do *not* let go with your left hand yet.

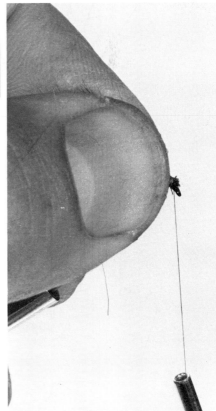

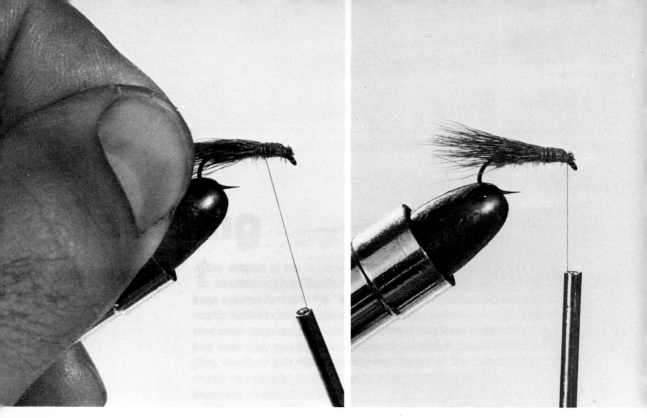

9 To prevent the hair from rolling on the hook shank, slide the fingers of your left hand back along the hair, and grasp the tips firmly. Now wind five or six loose turns of thread over the fibers towards the bend for a distance of one-third of the hook shank from the eye to the bend. As you do so, allow a few fibers to slip down along the side of the shank, to obtain a slightly rounded wing effect.

10 Repeat this process, returning to the eye; make two or three *more* tight turns right behind the eye to finally secure the wing. The loose winds are necessary to hold the deer hair together and prevent flaring. Coat the loose windings of thread with cement to prevent slippage.

Incidentally, if the body of the fly has not been constructed as earlier advised the deer hair will not lie flat on top of the shank.

11 Select one or two of the proper size hackle for the size hook you are tying on. Leave one quarter-inch of the stem bare and tie it in just behind the eye.

12 Make two or three loose turns over the stem portion to hold it down on top of the loose turns you made over the hair so that you can wind the hackle forward towards the eye. Then wind the thread back to the eye with two or three more loose turns. When you reach the eye, add several tight turns.

13 Wind the hackle forward and secure it behind the eye. Whip finish and head cement and your Solomon's Hairwing is complete.

ADULT CADDIS

sid neff hairwing

hook: Mustad 94840 or equivalent

thread: Fine, waxed; usually brown (color should add segmentation effect to body color)

body: Brown/grey fur dubbing (muskrat, etc.) or a dubbing mixture to match the natural

wing: Brown/grey deer hair or shades to match the natural

1 This pattern is tied in sizes twelve through twenty. Place a hook in your vise and spiral some fine brown thread onto the shank terminating at the bend. Spin a proper amount of muskrat underfur for the size fly you are tying onto the thread.

2 Once the dubbing has been spun onto the thread wind it forward along the shank terminating three-sixteenths of an inch from the eye of the hook.

3 Clip a section of brown/grey deer fibers from a piece of deer hide (approximately twenty fibers) and measure them along the shank so that they extend past the bend one-half again the length of the hook shank.

4 Tie in the deer hair fibers and clip the excess butts. The deer hair on this pattern *should* flare upwards off the shank of the hook very slightly.

5 Finish wrapping thread around the head, whip finish, and add a drop of head cement. The completed Sid Neff Hairwing should appear like this.

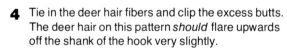

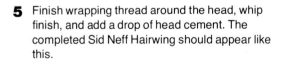

ADULT CADDIS

len wright
skittering
caddis

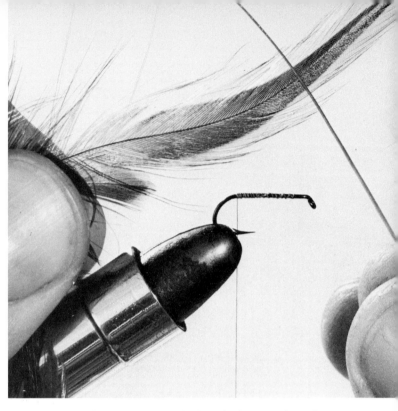

hook: Mustad 94840, 94833 or equivalent

thread: Fine

body: Soaked quill stem from large hackle feather or herl

wing: Very stiff hackle fibers, or mink tail guard hairs, to match coloration of the natural (usually gray)

hackle: Brown, or quality dry fly hackle of shade to match the natural

1 This pattern is tied on sizes twelve through twenty. Place a size twelve hook in your vise and spiral some fine thread (gray is good) onto the shank terminating at the bend. From a furnace rooster neck pluck one of the larger hackles from the base and pluck—do not strip—the fibers from the center stem.

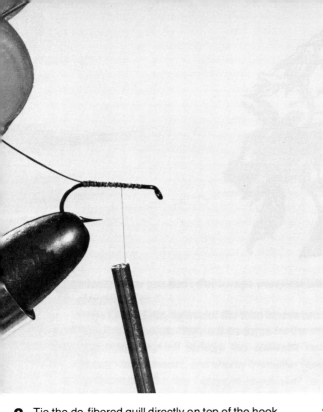

2 Tie the de-fibered quill directly on top of the hook shank by the tip end. It's best to soak the brittle stem in warm water prior to use, in order to prevent splitting. You can store presoaked stems in a small bottle. Len Wright himself ties this pattern with peacock or ostrich herl, which also works well.

3 Wind the thread forward to a point approximately three-sixteenths of an inch before the eye of the hook. In connecting turns, wind the quill stem to the thread. Clip the excess stem butt.

ADULT CADDIS

4 The wing of this pattern calls for very stiff (and long) spade hackle fibers in a dark grey shade. Most fly tyers have difficulty in obtaining this type of hackle. If you cannot locate such hackle, by all means use mink tail guard hair fibers. They are just as stiff and are actually more durable.

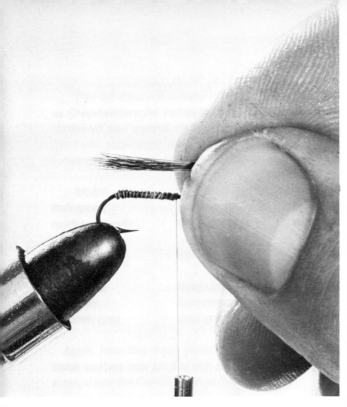

5 Cut a section of approximately ten fibers, from either a rooster feather or a mink tail, and measure them along the shank of the hook so that they extend past the bend for approximately one-half the length of the hook shank.

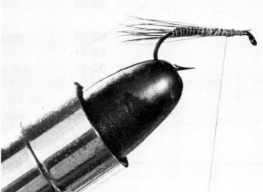

6 Tie this section of fibers in and against the far side of the hook shank.

7 Measure a similar quantity of fibers against those previously tied in, and tie them in and against the near side of the hook shank. In both cases clip the excess butts.

ADULT CADDIS

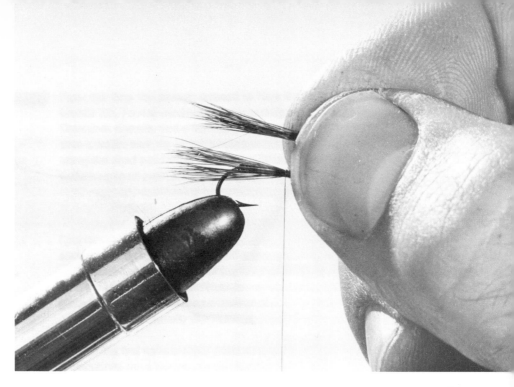

8 Measure a third section of fibers along the top of the hook shank so that the tips align with the first two sections. Tie this last section of fibers directly on top of the hook shank. Clip the excess butts.

9 As you can see from the photographs, all butt ends of the fibers extend to the eye of the hook. The hackle collar, which is next, will be wound over the butt ends of the fibers. Pluck a hackle feather from a rooster neck and measure the fiber diameter so that when the hackle is tied in it will form a collar similar in size to that of a conventional dry fly of the size you are tying. Trim the base of the feather and tie it in.

10 Wind the hackle collar, forming your hackle. Clip the excess tip. Whip-finish and apply a touch of head cement. The Len Wright Skittering Caddis is complete.

ADULT CADDIS

flat wing caddis

hook: Mustad 94840 or equivalent

thread: Fine, waxed

body: Grey dubbing (muskrat, etc.) or dubbing shade to match natural

rib: Brown hackle palmered and trimmed top and bottom

wing: Brown lacquered partridge feather

hackle: Brown or proper shade to match legs of natural, trimmed on bottom

1 The Flat Wing Caddis is designed to ride flush in the surface film. It is tied in sizes twelve through twenty. Place a hook in your vise and spiral some fine brown thread onto the shank terminating at the bend. Pluck a hackle feather from your rooster neck. Tie the hackle feather in at the bend and by the tip.

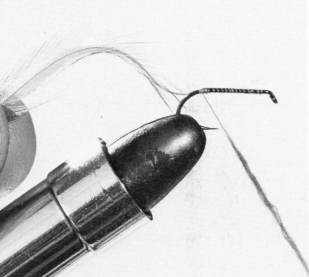 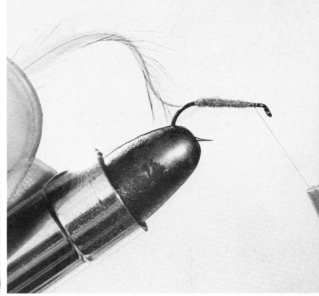

2 Spin some dubbing onto your thread.

3 Wind the dubbed thread around and along the hook shank to a point approximately one-eighth of an inch from the eye of the hook.

4 Go back and grasp the hackle feather with a pair of hackle pliers and wind it in an open spiral through the body and to the thread. Tie it down with the thread and clip the excess.

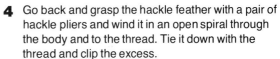

ADULT CADDIS

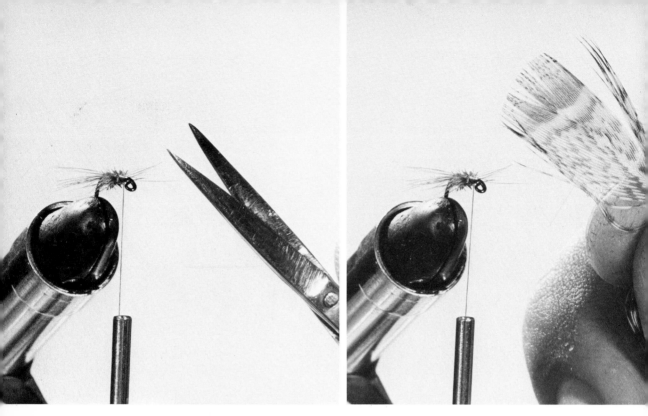

5 With a pair of scissors, trim the hackle fibers from both the top and the bottom of the hook shank. Only those fibers extending from the sides of the shank should remain.

6 Select a feather that is either mottled or solid depending on the wing pattern of the natural. It can be any of a variety of feathers: grouse, pheasant (all kinds), and woodcock all have such feathers.

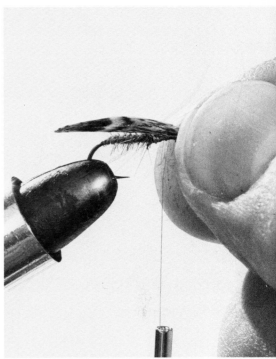

7 Separate the fibers from your feather so that only the tip section remains. It should be approximately three-sixteenths of an inch wide. The length of the section, when tied in, should extend slightly past the bend of the hook. Take the feather and paint it with vinyl cement. Allow it to dry. If you are doing a number of this type of pattern, it is best to prepare a quantity of lacquered feather sections in advance.

8 When it has dried, lay the feather section on top of the hook shank and measure it so that the tip extends slightly past the bend of the hook.

9 Tie it in just behind the eye, making sure that the feather lies flat on top of the hook. Clip the excess butts at the eye of the hook.

ADULT CADDIS

10 A top view.

11 Pluck another hackle, trim the base, and tie it in just behind the eye.

12 After the hackle has been wound, clip the excess
tip, whip-finish and add a touch of head cement.
With a pair of scissors trim all the hackles on the
bottom of the fly.

13 The completed Flat Wing Caddis should appear like
this.

ADULT CADDIS

delta wing
caddis

hook: Mustad 94840 or equivalent

thread: Olive, fine

body: Light olive fur (mink, etc.), or the proper shade dubbing to match the natural.

wing: Two grey hen hackle tips, or the color to match the natural

hackle: Brown, or any quality dry fly hackle to match the legs of the natural

1 This pattern is tied in sizes ten through twenty-two. Spiral the thread onto the shank of the hook terminating at the bend. Spin some fur dubbing onto the thread.

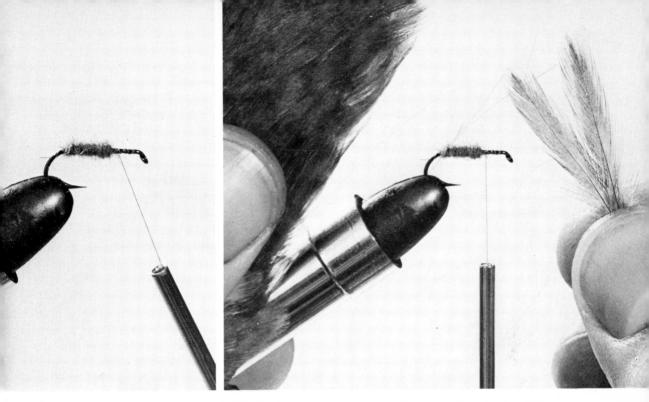

2 Wind the dubbed thread forward around the hook shank for approximately two-thirds its distance to the eye of the hook.

3 From a natural hen neck (or a dyed neck if the natural is not available) pluck two of the feathers. If you can obtain an Indian hen neck, all the better, since these chickens seem to produce small yet compact feathers which are perfectly suited for this type of fly.

4 Measure one of the hen hackle tips along the body of the fly. The tip of the hackle should extend slightly past the bend of the hook as it slants outward at a forty-five-degree angle. Note the point at which the hackle tip feather is to be tied to the shank of the hook, at the end of the dubbing.

ADULT CADDIS

5 Strip off the remaining butt fibers from the hackle tips and tie it in on the far side of the hook shank.

6 Measure a second hackle tip feather so that it will be of the same size as the first but will protrude from the opposite side of the shank.

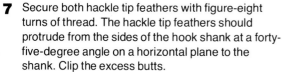

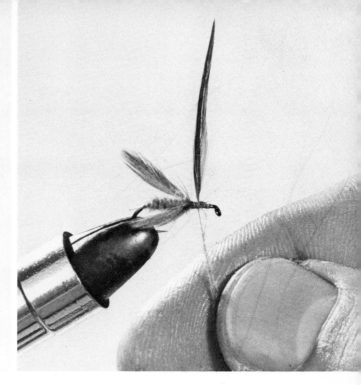

7 Secure both hackle tip feathers with figure-eight turns of thread. The hackle tip feathers should protrude from the sides of the hook shank at a forty-five-degree angle on a horizontal plane to the shank. Clip the excess butts.

8 Pluck a feather from a rooster neck; select a proper size feather for the hook size you are using. Trim the excess fibers from the butt and tie it in at a midpoint between the wings of the fly and the eye of the hook.

Forget the hackle feather for now. You can prop it straight up by taking a few turns of thread around the base.

Spin some more dubbing onto your thread.

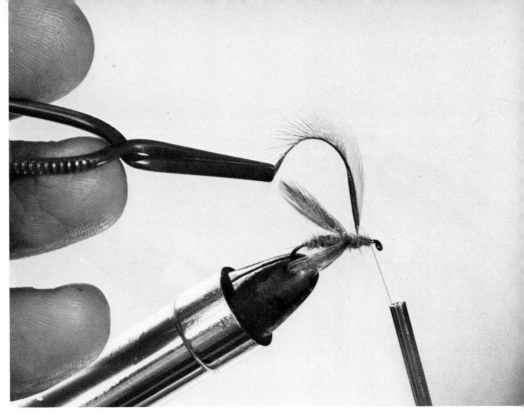

9 Wind the dubbed thread around the shank of the hook and fill in the space from the wings to the eye of the hook. The dubbing should form a natural continuation of the body of the fly. It should be worked around the protruding hackle feather standing erect. Once the dubbing has been worked up to the eye of the hook, grasp the hackle by its tip with a pair of pliers and wind the hackle collar.

Whip-finish the fly, cut your thread away, and add a touch of head cement to the windings.

10 Take a pair of scissors and trim the hackles from the bottom of the fly.

11 The completed Delta Wing Caddis should look like this from the side.

12 A top view.

ADULT CADDIS

hare's ear
or vermont
caddis

hook: Mustad 94840
or equivalent

thread: Grey, fine

tail: Grizzly hackles tied short
(optional)

body: Hare's ear dubbing

hackle: Brown and grizzly mixed

1 This pattern is tied in sizes ten through twenty.
Place a hook in your vise and spiral some fine grey
thread onto the shank terminating at the bend.

    The Hare's Ear is most often tied without a tail;
however since some people like to add a tail for
stability (caddis do not have an actual tail), this step
is optional. Pluck a few fibers from a grizzly hackle
feather and measure them for the tail. The tail
should not be as long as one you would tie for a
mayfly imitation.

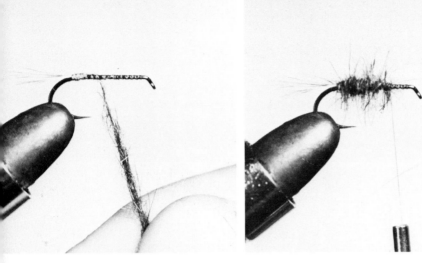

2 From an English hare's ear or mask, pluck or cut some of the coarse and soft dubbing, spinning it onto your thread. Since this dubbing is extracoarse you may want to wet your fingertips slightly.

3 Wind the dubbed thread forward about three-quarters of the length of the shank to the eye, thus forming the body.

4 Tie in one brown and one grizzly hackle feather by the butts. Wind them one at a time around the shank forming the hackle collar.

5

Whip-finish a head on the fly, apply a touch of head cement and your completed Hare's Ear Caddis should resemble the one in the photo.

quill wing caddis

hook: Mustad 94840, 94833 or equivalent

thread: Grey, fine

body: Light olive dubbing or dubbing to match the natural

wing: Two sections from paired flight feathers (mallard, turkey, etc.)

hackle: Dark ginger, or quality dry fly hackle to match the legs of the natural

note: When imitating species with pronounced mottled wings, substitute sections from turkey wing quills or equivalent.

1 Place a hook in your vise and spiral your thread onto the shank terminating at the bend.

Spin a fair amount of dubbing onto the thread (mink is just fine).

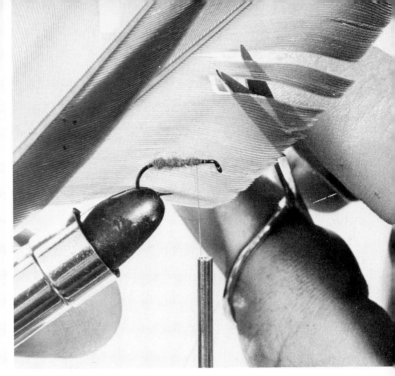

2 Wind the dubbed thread along the shank for about three quarters of the distance to the eye.

3 From a matched pair (right and left) of flight feathers clip respective quill sections measuring approximately one-eighth inch in diameter.

4 With a dubbing needle, lacquer the inside of the quill section so that the fibers do not split. Vinyl cement is excellent for this since it is fast-drying.

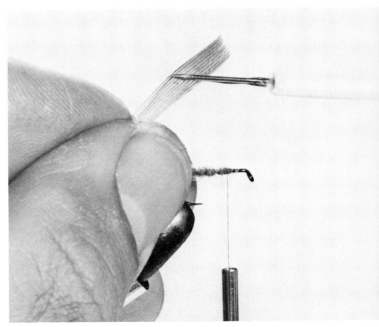

ADULT CADDIS

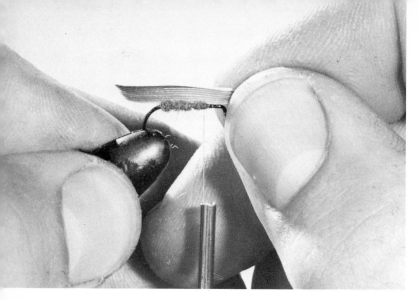

5 Place one of the sections on the far side of the hook shank and measure it so that the tips extend approximately half of the length of the shank past the bend of the hook.

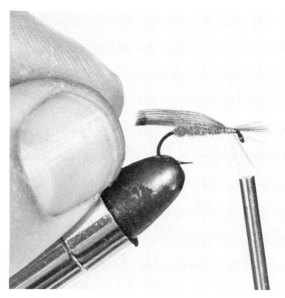

6 Tie down the quill section on the far side of the hook.

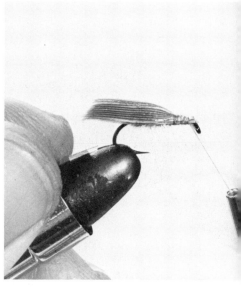

7 Measure the other quill section against the near side of the hook and align the tips of the quill with the one previously tied in. Tie in the near section and clip the excess butts protruding out over the eye of the hook on both sections.

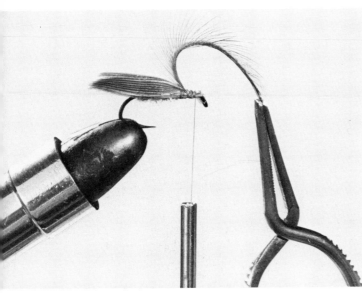

8 Pluck a hackle feather from a rooster neck, tie it in, and wind it around the shank for a hackle collar.

9 Whip-finish and apply a touch of head cement. The completed Quill Wing Caddis should appear as it does here.

ADULT CADDIS

clipped hackle body

The following is a method created, as far as we know, by Walt Dette of Roscoe, New York, to form clipped hackle bodies where they are called for. This technique is especially useful for large flies when the required amount of dubbing or other material would absorb water too readily, and thus restrict the floating quality of the artificial.

This method also allows you to use some of those hackles from a rooster neck which are rarely ever wanted, since they are too large for conventional hackling. Also, the hackles with thin fibers and thin center quills which are not very good for dry fly hackle work very well for a clipped body.

The hackle can be any shade. Your color selection should be for the natural insect you are imitating. In this instance we have used a furnace rooster hackle.

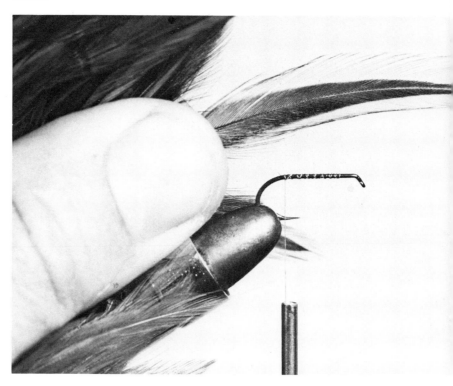

1 Large furnace rooster hackle prior to preparation

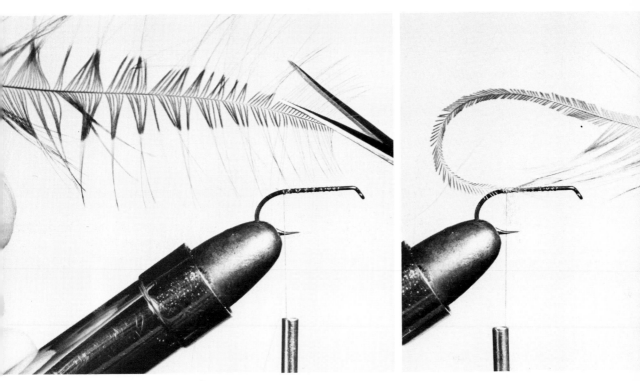

2 Hackle fibers have been stroked rearward and extended at right angles to the stem. Trim the barbules as close to the stem as desired.

3 The tip or butt of trimmed hackle is tied to shank of hook. Although we illustrate tying in the tip of the feather, we find tying in and winding from the butt seems to give a more consistently uniform body.

4 Hackle wound to hook and secured, thus forming clipped hackle body. The fibers may be trimmed at this time for uniformity.

adult caddis: points to remember

1 Present your fly properly; your imitation should appear as natural to the trout as possible. By fishing with a length of line you can control, most of your difficulties will be solved.

2 Fish the fluttering type of caddis imitation downstream, or across and downstream. Work the fly upstream in very short movements, simulating the natural trying to get off the surface. (Keep the tip of the rod high and the line as short as possible for the situation.)

3 Use the upstream hold, or extension cast when required, as opposed to mending line in order to obtain a drag-free float.

4 Observe the form of the rise. Splashy rises are usually indicative of a caddis hatch, or one that is about to take place. Gentle head-and-tail rises will indicate (especially during a flush hatch), that there are many crippled, dead, or troubled caddis flies on the surface; the odds are also more favorable for larger fish to be present and feeding.

5 If the presence of caddis is evident, but trout are selective, check to see whether there is more than one species coming off. Adjust to the preference of the trout.

6 If, during a flush hatch, the conventional downwing patterns do not seem to work as well, try using a Delta Wing. At the other extreme, when there is a slack period, also experiment with the Delta Wing, since it will serve as a multi-purposed fly imitating not only the caddis, but spent mayflies and terrestrials as well.

hook to insect size correlation

Measurements are shown in millimeters for maximum accuracy. This chart is used to show the relationship of the average length of the natural insect (from head to wing tip) and the hook size that the imitation should be tied on. For example, if you had collected a specimen that measured 14 mm from head to wing tip, you would refer to the chart and see that the imitation should be tied on a #16 hook. In most instances the wing length will be about one and one half times the length of the body. However, in many instances with larger flies above 13 to 15 mm, the wing may be as much as twice the body length.

INSECT SIZE mm (head to wing tip)	BODY SIZE mm	HOOK SIZE
6–7	4½–5	22
8–9	5½–6	20
10–11	6½–7	18
12–14	7½–8	16
15–17	8½–9½	14
18–20	10–11	12
21–24	12–13	10
25–29	14–15	8
30–34	16–18	6

The hook illustrated is Mustad #94840 (#94845 if barbless hook is desired). In sizes #20 and #22, hook #94859 (straight eye) is suggested for maximum hooking ability.

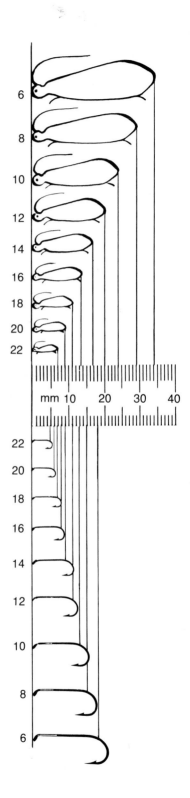

ADULT CADDIS

ovipositing

the final curtain

The subject of the mating rituals and spinner falls of the mayfly has been covered very comprehensively in angling literature. Many patterns have been created to match the "fall" of the hatch as well as its emergence. In the mayfly families spinner falls are common; most of us have witnessed an occasion when large quantities of dying or dead mayflies were on the water, where they were easy prey for trout. On the other hand, there is little if anything in angling literature that discusses the importance of the ovipositing (egg-laying) caddis as a food for trout; it has almost been overlooked. And yet it is extremely important, as we shall see.

Since the caddis lives as an adult for a generally much longer period of time than the mayfly, his appearance on or over the stream (other than during the emergence or ovipositing period) is quite common. Unlike mayflies, which go through a dramatic physiological change during emergence and ovipositing, from the emerged adult (subimago) to the mating adult (imago), the caddis retains the same basic form and coloration through all of its adult phase. For this reason alone, it is difficult for the angler to know when caddis are ovipositing. Also, when trout are feeding on spent ovipositing mayflies, the dying insects are on the water in such a quantity that the angler can recognize what is taking place. The caddis life cycle does not end in this manner. Rarely will you see large quantities or even smaller quantities of spent caddis flies on the water. And the reason it is rarely seen is because it rarely happens.

After laying its eggs the caddis, in most instances, leaves the water and returns to the land. There are occasions when caddis will mate two or even three times before expiring. They do not, moreover, succumb to the rigors of mating and egg-laying, as in the case of the mayfly.

What then, is the value of the ovipositing caddis, if it does not collapse completely spent on the water, as the mayflies do, so trout can feed on it?

There are two reasons, one very good in itself, the other spectacular. Firstly, most caddis females do expose themselves to the trout in the act of ovipositing in some way, most often by flying close to the surface, dipping their abdomens into the water to release egg masses into the stream. And, as we have already mentioned, they may do this more than once. More importantly, in some families, the female actually *enters* the water, sometimes making its way to the bottom to deposit its eggs, and *re-emerges!* Obviously, this behavior has significance for the angler, as we'll see, in several ways. Caddis actually oviposit in four different ways and can be listed in groups as follows:

FAMILY	OVIPOSITING CHARACTERISTICS
A Rhyacophilidae Philopotamidae Hydropsychidae Hydroptilidae Polycentropidae	Females of these families enter the water and lay strings of eggs on the bottom, usually on rocks. They sometimes crawl down stems of plants or swim to the bottom.
B Leptoceridae Phryganidae Brachycentridae Odonticeridae	Females carry egg ball masses (similar to mayflies) and dip their abdomens into the water; or enter the water entirely and release the egg masses which attach themselves to a submerged object (stones, rocks, and the like).
C Helicopsychidae Goeridae Lepidostomatidae	Egg masses are deposited in the water, or in damp areas near the water.
D Limnephilidae	This family deposits eggs above the water on stones or plants, or sometimes on twigs of trees above the stream. Rains eventually wash the larvae into the stream.

Obviously the species of Limnephilidae, group D, are of no value as food during ovipositing since they do not return to the water to perform this function.

The families in groups B and C both carry egg ball masses and deposit their eggs much in the manner of a mayfly by dipping their abdomens into the water, and sometimes even going slightly below the surface. When ovipositing is completed they leave the water and return to land. They are available to the trout for a short period of time, though they do seem to linger a bit longer than during an emergence period. This apparent slowness may be due to diminishing energy.

The species of the families in groups B and C may be imitated with the various adult caddis patterns. Especially effective are the Delta Wing and the Flat Wing imitations. The procedure for fishing the fly is about the same as that used during the emergence period, although a bit more "dead drift" technique may be employed. The rise form, incidentally, will not be as sharp or splashy.

In dealing with the families in group A we have a startlingly different situation. This discovery, for me, began with several puzzling experiences astream, when I was fishing a single hatch of caddis that I was sure I had identified. The fish were obviously rising, but neither a closely matched pupa or dry fly seemed to work. Evidently, the trout were seeing something I could not. Then I read an article in *The Canadian Entomologist* by Philip S. Corbet which related the results of research conducted on the Saint Lawrence River on the behavior pattern of caddis during emergence and ovipositing, which had real significance for the angler as well as the entomologist.*

Tests were conducted by mooring two float traps over caddis waters, each about three feet in diameter. At regular intervals, the emerging caddis that were caught were removed from the traps.

The traps were changed six times a day at four-hour intervals. This test proved that emergence usually occurred between sunset and midnight, with some variation from day to day, depending on water and air temperature. The most interesting discovery, however, was that upon inspection and dissection of captured insects, it was noted that many of the insects captured were adult females that had just returned from ovipositing. As mentioned earlier, the females of the species in group A pass through the surface to lay their eggs on the bottom, returning to land after their task has been completed.

An even more interesting fact was that the schedule of the returning ovipositing species was quite consistent with the period of emergence of the same species. That is, the ovipositing females returned during a hatch of pupae. Now this was news! It gave me the answer to the frustrating experiences of the past. If the emerging insect and the ovipositing insect appear the same to the eye, and if the ovipositing female often comes back out of the water during the emergence of the same species, how can one tell which is which? We don't! You and I will not be able to tell the difference, that is. The trout, however, can, because prior to breaking through the surface film each stage (the emerging pupa and the returning ovipositing female) will have a different silhouette as it comes up through the water. The emerging pupa will have short, stubby wings, cased in the shuck, half the size of its body and slanting downward, while the returning female will have full mature wings, longer than the body and extending along the top.

Now where does that leave us, unless we don some scuba gear and go under with a magnifier? The answer is that we have to experiment. If you are not getting results using the conventional methods of fishing the pupa, or the

Canadian Entomologist 98 (October 1966): 1025–34.

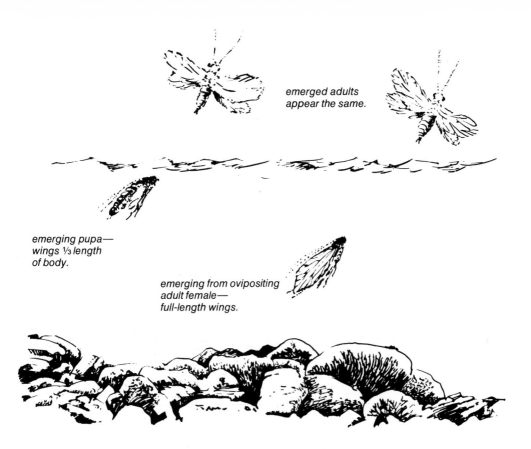

emerged adults
appear the same.

emerging pupa—
wings ⅓ length
of body.

emerging from ovipositing
adult female—
full-length wings.

adult caddis during a hatch, *try imitating the returning female with a wet fly pattern.*

The first time I encountered this phenomenon personally was during an evening's fishing on the Beaverkill in New York. A mating flight of *Psilotreta labida,* which return in great quantities, was in progress. There was little evidence of activity since few insects were on the water. As dusk approached, I noticed similar insects emerging from the surface. Shortly thereafter trout began to rise to what I thought was the emerging *Psilotreta.* After many casts but only one fish, I decided to do a little collecting. While bending over in the stream I noticed several small slightly different caddis flies on my wader, just above the water line. The insect was familiar to me. I selected a hairwing pattern which imitated it, still without results.

Then the thought struck me. Were these the returning flies of the same species? I switched to a Delta Wing which was a close imitation, forced the wings back so that they would stay close to the body, and fished the pattern as a wet fly, or emerging pupa. The results were astonishing: four fat brown trout were taken and released within twenty minutes. The trout were rising sharply, indicating that the insect was fairly rapid in its return. I collected some samples of the species and was not surprised when I learned they belonged to the family Rhyacophilidae (group A).

This type of occurrence has repeated itself a number of times since that day. I have since also tied imitations for these group A species that work quite well. They are nothing revolutionary, simply a wet fly without a tail. The color of all the materials used on the fly should, of course, be representative of the adult natural. The wing should be made of materials such as hen hackle points or soft wing quill sections, such as starling, crow, grouse, partridge and the like.

One of the keys to the situation is being observant and experimenting with your fly selections. The correct size of the pattern is important for best results. Here again is a good time to try a cast of two flies, one at the tip and one as a dropper. Using an arrangement of one emerging pupa imitation and one returning wet fly on the dropper, or perhaps a dry fly with a wet fly dropper may just turn the trick. If you take fish consistently on any one of the patterns, that's your key.

Here is another idea that may work well in relation to ovipositing: if a particular caddis species has emerged or oviposited during the night, fish a Delta Wing dead drift very early the following morning, before any hatches begin. I have consistently had quality fishing using this method. Whether there are some dead insects remaining from the emergence or ovipositing period of the night before I have not been able to verify, but the results of fishing the Delta Wing have proven it a worthwhile tactic. At this time the rises are always gentle and deliberate.

In doing the research for this volume I have spoken to many anglers in every state where trout abound, and it is worth noting here, that in many areas, such standard wet flies as the Gold-Ribbed Hare's Ear and the Leadwing Coachman are used as imitations for the caddis pupa during emergence periods. However, though fished as pupae, they are many times taken by the trout for a female caddis from family group A returning from ovipositing. The use of wet flies as pupal imitations is not regional. Anglers in New Hampshire, Pennsylvania, Michigan, and Colorado, for example, have listed the exact same patterns. Wet flies, in general, are taken by trout not as spent, dead, or emerging mayflies, but as food in the form of caddis pupae and ovipositing adults. Perhaps this is why the old standard wet flies are still with us. They serve their purpose, though most of us did not know what that purpose was for many years.

ovipositing: points to remember

1 Species of certain families of caddis enter the water to lay their eggs and return very close in time to the emergence period of the same species.

2 Delta Wings and wet flies without tails are good imitations of the returning female caddis.

3 Delta Wing patterns usually work very well early in the morning.

caddis cousins

"**i** certainly hope you don't discourage everyone with an unpronounceable list of Latin names."

It was Ernie Maltz, calling for the third time in two weeks.

"No, Ernie," I said, "they will be listed for the sake of accuracy in description, but the listing will include the common names."

"Good," he answered, and shortly thereafter hung up.

For the rest of the day I thought about what he had said in relation to this particular chapter. I had a problem: *there are no common names associated with caddis*—at least not in the same manner a common name is associated with a mayfly. Certainly such caddis names as grannom, shadfly, dark sedge, and the like could be referred to as common names, but even they were identified with many different species, depending on the region. On the other hand, in the category of mayflies, any angler could relate the fly to one insect species by using such names as quill gordon, hendrickson, light cahill, blue-winged olive, or sulfur dun. Mayflies have been familiar to anglers for generations, and the association between specific naturals and the flies named after them has been taken for granted. Caddis flies have existed in our streams just as long as mayflies, but it is only recently that they have been given their due respect as trout food.

So, how do you refer to the common names of caddis when there are none? At this point in time you cannot say, "There's a Henryville coming off" or "A Jorgensen is due to hatch shortly!" or "I checked the stomach contents and found it loaded with Solomons." Perhaps one day—but for now, it would never do. And besides, the few caddis patterns that have become widely known are tied in a variety of colors to match the shades of different species. So that's where it's at—color! The only ground for a common name for caddis is the color common to the particular species. Thus, a green/brown caddis is one with a green body and a brown wing. So we have a usable system of names.

Still, in the emergence charts that follow in this chapter you will find many different species of caddis flies that are identical in color, and there seems to be no strict way to identify caddis species by color or emergence time as there is for mayflies.

A hendrickson, for example, is a hendrickson *(Ephemerella subvaria).* It comes off in May, it has three tails, brownish/grey wings, a pink/grey body and you tie it on a size fourteen. And, though it may begin hatching earlier or later during its season, depending on the region, once the major emergence has taken place it does not repeat its performance until the following year. The quill gordons, hendrickson and march brown are equally dependable (though not always reasonable), and their distinctive coloration makes them easily recognizable to the angler as a species.

The caddis fly, with a few exceptions, is much more confusing. Look-alike species will emerge throughout the season. However, if you become familiar with caddis characteristics and habits as *families,* the only other factors you really have to be concerned with are their size and color. Color will sometimes vary in families or even genera but these exceptions can be isolated for a given area. In a way it's a blessing in disguise, especially when imitating the adult stage. You don't have to know the species to match the hatch. Knowing the species, or at least the genus, *is* more important when fishing the larval and pupal stages of the caddis—to know which ones are free-living and which are cased. Identifying the families present in your local streams is the first step in fishing the caddis.

This chapter has been broken down into four sections of emergence information covering the caddis most commonly found in the regions important to the fly fisherman. The emergence tables listed at the end of each section are designed for ready reference to the insects most prevalent in each region. You'll notice that many families overlap, and in many instances are prevalent in all four regions.

Remember, these emergence tables are only general; seasonal times and daily emergences will vary from area to area and stream to stream. Air and water temperature will also affect emergence. The charts are really designed to give you an idea of what caddis activity goes on in your region, and if you do locate the genus of insects mentioned, to give you a ready-made idea of size and color.

Patterns for the emerging insects have not been listed next to the genus represented because there would be too much confusion. Consider especially that there are many styles of patterns; some directly imitative, some impressionistic, some fished in fast water, some in slow, and a number of other variables.

The most obvious solution to matching the hatch would be to have an imitation for every insect that you are likely to encounter. Many ardent fly fishers do, this one included. For many anglers, however, especially the novice and the nontiers, this level of preparation could be a monumental task. Therefore, in an attempt to simplify the problem a bit, we have listed below six variations of color in several sizes which can represent about seventy-five percent of the naturals that will be encountered across the country.

This list does not include imitations of certain important specimens, such as *Cheumatopsyche campyla* (#22) in the East, *Dicosmoecus* (#6–8) in the Northwest and several others which may be necessary for selective hatches. But if you don't want to get totally involved at first, this list will give you a basic selection to which you can add for local areas.

BODY	WING	LEGS	SIZES
dark green	brown grey	bronze dun	14–16
medium greyish olive	light brownish grey	rusty dun	16–20
light yellowish olive	tan	light ginger	16–18
brown grey	brownish grey	medium brown	16–20
brownish	light brown	ginger	14–18
charcoal grey	dark grey	dark bronze dun	14–20

Again, note that these patterns are for the dry fly ties, but the pupa of the same species may be tied with slightly darker wings. For larva coloration and sizes, check the Caddis Families in Chapter 1.

The column pertaining to size of the genus represented is for the overall insect size, from head to wing tip. Hook size is for the body length only. There is a ruler on the market called The English-Metric Scale Rule, which designates millimeters and relates them to inches. It is small, light and handy. It can be carried in a fishing vest pocket in addition to being used in the home. When hook sizes are stated, I have used the Mustad Model 94840 as the standard. See the insect to hook size correlation chart in Chapter 4 for verification of actual specimens.

The Genus/Species column will in most cases list the genera only. Only where a specific species is important to the angler will it be listed. The idea of the charts is to provide some basis for fly tying reference and general understanding—not add confusion, which in the case of caddis is not difficult to do.

Many trips were made to farther reaches of caddis country; many more calls to friends in every important state were made, and so many samples were sent to Dr. Oliver Flint, Jr. in Washington, an eminent caddis entomologist at the Smithsonian, that I must believe his patience with fishermen had to be reaching the breaking point with each shipment or telephone call. But his own interest in caddis must have outweighed everything else, and in one instance the authors were surprised to learn that one of their collected specimens was a species not yet recorded in that state by Dr. Flint.

As information filtered back from the Smithsonian, identification became easier, even for us. Here is the benefit of our efforts, and Dr. Flint's help, in the form of regional emergence charts for the caddis families significant to anglers.

the east From the time the season opened in New York (April 1) until it closed (September 30), I overturned more rocks and stones in the Beaverkill, Willowemoc, Delaware, Esopus, and a number of smaller streams, than I would care to load onto a trailer van. As I worked in the water now and again a car would stop along the road paralleling the stream. The occupants would watch and stare, wondering in all probability what a grown man was doing stumbling and scurrying through the fast-flowing water chasing an unseen something with a three-inch butterfly net.

Travels through Connecticut and Massachusetts added more information. Specimens in bottles arrived by way of United Parcel from John Harder, who did some collecting for the authors in Rhode Island.

Pennsylvania, the home of the Henryville Special (the most popular caddis fly in the United States), and New Jersey abounded not only in caddis but in anglers who knew their value as a trout food.

The East, especially the Northeast, has the largest caddis populations in the country.

For the first few weeks of the trout season in the early spring, bright green *Rhyacophila* larva worms can be found in trout stomachs (usually a size #16). If you look under rocks you will also find the largest of the "Rhyacs" *(Rhyacophila fuscula)*; they are bright green, approximately 15 to 18 millimeters in length. These are the larvae so well imitated by the latex caddis larva pattern.

The first significant adult caddis appear in late April and early May, depending on the climate of the region. Generally, the more northern latitudes have a later emergence. The dark grey caddis are of the genera *Chimarra* and *Dolophilodes* of the Philopotamidae family. Although they are only 7 to 9 millimeters (about size 18 or 20 as an artificial), trout will feed on them selectively as they emerge during the midday hours.

When the hendrickson mayfly *(Ephemerella subvaria)* has almost completed its annual appearance, one of the major caddis of the season emerges. The species is *Brachycentrus numerosus,* commonly referred to as the grannom caddis, or known locally in the Catskill region as the shad fly. These insects emerge in great quantities, often in the morning hours. Their massive numbers in mating flights upriver are a spectacular sight. However, most often they do not oviposit until sometime later, in most instances after dark. From the inception of the *numerosus* emergence and for a three-week period thereafter, the emergence of two or three other *Brachycentrus* species occurs. Their size varies from 11 to 15 millimeters with some slight differences in color shading.

From the beginning to the end of the season, various species of the *Hydropsyche* genus (family Hydropsychidae) will make their appearance. These are the most common of all the caddis on this continent. They are easily recognized by pronounced triangular-shaped wings often fringed with dash-type markings around the trailing edges of the wing. Trout seem to key on these species when they emerge. Although the coloration is quite similar throughout the various species with sizes ranging from 7 to 11 millimeters, the trout have

158

been quite selective at times, especially in flat water conditions.

The most numerous hatch in the East occurs from beginning to late June, again depending on the location and water temperature; it is the genus *Psilotreta* of the Odontoceridae family. This genus is unique to the East. The species *Psilotreta labida* is the most common, inhabiting medium-to-large-size streams and rivers. Less abundant is *Psilotreta frontalis,* which is found in the smaller mountain streams.

The tiny microcaddis plays an important role at times, especially during the summer months when most of the significant insect populations have already emerged and the fish are searching for food. It is present, however, from the beginning to the end of the season in size 4 to 7 millimeters (hook size 20 to 24). I have observed trout feeding selectively on microcaddis even though larger species were available. One particular species, *Cheumatopsyche campyla,* of the Hydropsychidae family, brings that type of response. It measures 6 to 7 millimeters (most effective on a size 22 hook), with a bright kelly green body, slate grey wings, and rusty dun legs. (Unless you hold a specimen in your hands you will not be able to determine its color.) If your imitation is correctly tied, trout will accept it eagerly. This species often emerges when there are other insects present and often goes unnoticed by most anglers who are trying to understand the reason trout are not taking their more obvious offering.

Along with the *Psilotreta,* species of the Rhyacophilidae family are also emerging simultaneously. Since certain *Rhyacophila* species have dark wings, as do *Psilotreta,* they are easily mistaken for the latter. In the hand, the angler can see the body color difference as well as slight difference in size. This difference of species is very important at time of ovipositing, since *Rhyacophila* are a group A genus which enters the water to deposit its eggs on the bottom.

The tiny dome-shaped pebble cases attached to the sides of rocks (these casemakers are not mobile as larvae) belong to the genus *Glossosoma,* of the Glossosomatidae family. *Glossosoma nigrior* most commonly emerges in June and July. It is approximately 8 to 9 millimeters long and generally brown in coloration. (Glossosomatidae are closely related to Rhyacophilidae in all respects that are important to the angler.)

Macronemum zebratum, a species of the largest eastern genus of the Hydropsychidae family emerges in late May through June. It inhabits larger streams and rivers and is easily recognized by its very long antennae and pronounced brown/cream mottled wing pattern.

Species of the large caddis from the Phryganidae family emerge in July and August from the slower waters and ponds of the New England states up to Canada. Most emergence takes place during darkness.

In late summer and early fall, some members of the widely varied Limnephilidae family appear. Most common is the species *Pycnopsyche guttifer.* Most anglers have seen its larval stick-case, which measures over an inch.

The *Hydropsyches* continue into the fall, while *Dolophilodes distinctus* of the Philopotamidae family emerges throughout the winter. The wingless female of *Dolophilodes distinctus* is often seen on snowbanks on a warm winter day.

emergence table—east

	CLASSIFICATION	SEASON	TIME OF DAY
FAMILY Genus	Philopotamidae *Chimarra* *Dolophilodes*	April/early May	late morning/ afternoon
FAMILY Genus species:	Brachycentridae *Brachycentrus* *numerosus*	early/mid-May	early morning/ afternoon
species:	unidentified	early/mid-May	early morning/ afternoon
species:	*nigrisoma*	mid/late May	afternoon
FAMILY Genus species:	Hydropsychidae *Hydropsyche* *slossonae* *morosa* *recurvata* *bifida* *bronta*	early May/ September	sporadic
Genus species:	*Cheumatopsyche* *campyla*	beginning/ mid-June	sporadic
Genus species:	*Macronemum* *zebratum*	late May/June	late morning/ afternoon
FAMILY Genus species:	Rhyacophilidae *Rhyacophila* *carolina*	late May/early June	afternoon/ early evening
species:	*fuscula*	May/June	sporadic/ mostly evening
species:	unidentified	beginning/ late June	mainly evening

ADULT INSECT SIZE (head to wing tip)	COLORATION		HOOK SIZE (ADULT)
7–10 mm	BODY: WING: LEGS:	black medium grey dark bronze dun	#18–20
13–15 mm	BODY: WING: LEGS:	dark greyish olive brownish grey bronze dun	#14–16
13–15 mm	BODY: WING: LEGS:	pale apple green light tan light ginger	#14–16
11–12 mm	BODY: WING: LEGS:	greyish amber medium grey rusty dun	#16–18
8–13 mm	BODY: WING: LEGS:	greyish olive light brownish grey rusty dun	#16–20
8–13 mm	BODY: WING: LEGS:	greyish olive dark tan ginger	#16–20
6–7 mm	BODY: WING: LEGS:	bright kelly green medium grey rusty dun	#22
18–20 mm	BODY: WING: LEGS:	Dark brownish grey mottled brown tan dark brown	#12–14
10–12 mm	BODY: WING: LEGS:	greyish brown brownish grey medium brown	#16–18
12–14 mm	BODY: WING: LEGS:	dark green grey brown mottled rusty dun	#14–16
11–13 mm	BODY: WING: LEGS:	greyish rusty cream brownish grey rusty dun	#16

continued

CADDIS COUSINS

emergence table—east *continued*

	CLASSIFICATION	SEASON	TIME OF DAY
FAMILY Genus species:	Glossosomatidae *Glossosoma* *nigrior*	late June/ August	sporadic
FAMILY Genus species:	Odontoceridae *Psilotreta* *labida* *frontalis*	beginning/ late June	mainly evening
FAMILY Genus	Leptoceridae *Oecetis*	late June/ August	morning/ early afternoon
FAMILY Genus species:	Helicopsychidae *Helicopsyche* *borealis*	June/September	sporadic
FAMILY Genus species:	Phryganeidae *Ptilostomis* *ocellifera*	July/August	dusk into dark
Genus species:	*Phryganea* *cinerea* (northern New England)	July/August	dark into dusk
FAMILY Genus species:	Limnephilidae *Pycnopsyche* *guttifer*	late August/ September	late evening to dusk
Genus	*Neophylax*	late August/ September	evening

ADULT INSECT SIZE (head to wing tip)	COLORATION		HOOK SIZE (ADULT)
8–10 mm	BODY: WING: LEGS:	brown greyish tan rusty dun	#18–20
13–15 mm	BODY: WING: LEGS:	dark green and charcoal mottled dark greyish dark brownish grey	#14–16
9–10 mm	BODY: WING: LEGS:	light brown mottled light brown dark ginger	#18–20
5–7 mm	BODY: WING: LEGS:	tannish grey brownish grey rusty dun	#22
18–22 mm	BODY: WING: LEGS:	light brown medium brown mottled dark ginger	#10–12
21–25 mm	BODY: WING: LEGS:	brown mottled tan/brown light brown	#8–10
18–20 mm	BODY: WING: LEGS:	brown mottled light brown light brown	#12
10–11 mm	BODY: WING: LEGS:	tan mottled tan ginger	#18

CADDIS COUSINS

the midwest (northern tier)

I have fished in Michigan and I am going back to Michigan to fish again. Michigan has it all: trout, steelhead, salmon (all kinds), and other gamefish— and mayflies, stoneflies, terrestrials, and caddis. It also has anglers who know all the fish, and all the insects. And the only time they are not trying to bring the two together is during the steelhead run. Then everything else—well, almost everything—is forgotten.

Geographically, Wisconsin is directly across the lake from Michigan's best trout water, and it offers all of the same type of fishing and all of the insects that are found in the famous Michigan streams. However, it is not as popularized, and I'm told by some local anglers they wish to keep it that way. These anglers do not work for the Chamber of Commerce.

Gary Borger lives in Wisconsin, in Wausau, to be exact, a professor of botany by college degree, an entomologist by self-decree. He is also an angler, a professional flycaster and a very creative flytyer. How he finds time for the flora I'll never know. When I first spoke to Gary concerning caddis flies in his state I learned that he had independently been doing his own research along the same lines. Here are some of his findings.

Several species of netmakers of the genus *Chimarra* (family Philopotamidae) begin to emerge in the Midwest in April and May; these are the famous "finger nets." Smaller hatches will take place throughout the season, mostly in fast-flowing water. The larvae are approximately 10 to 12 millimeters long, off-white in color. Adults are dark brown to black, measuring approximately 7 to 9 millimeters in size from head to wing tip.

Scarcely has this hatch reached its peak when the genus *Rhyacophila,* called the "green rock worm" in its larval form, begins to emerge here, almost on schedule with its Eastern relatives. Remember, these larvae are completely free-living, and bright green in color, varying from 8 millimeters in the smaller species to 15 in the larger *fuscula*.

Again, this is the caddis that is best imitated in larval form by the use of the latex body, since there is little evidence of gilling present on the body. The adult, which measures about 12 millimeters, emerges in May. It has slate grey wings, a greenish body, and fairly short antennae for caddis. Females oviposit by entering the water and laying their eggs on submerged objects.

In May and early June several species of the genus *Brachycentrus,* family Brachycentridae, begin to emerge. Commonly called the "grannom," it is one of the most widely known of all caddis, and inspection of Midwestern streams showed the chimney cases of this genus in great abundance. Adults are dark brown and about 9 millimeters in length. Females carry the familiar bright green egg sac, which they dip into the water during ovipositing.

Next come several species of the genus *Mystacides,* family Leptoceridae. The adult is blue/black, approximately 7 to 9 millimeters in length, with exceptionally long antennae (14 to 18 millimeters). It is commonly called the "black dancer." Emergence is sporadic during May and the early parts of the season, becoming much more concentrated in August and September.

During the night and early morning hours in June and July the genus *Ptilostomis* (several species), family Phryganeidae, will begin to emerge. Primarily a nocturnal hatcher, the "orange sedge" is one of the local Midwestern names given to these large caddis flies since they are of a rusty orange color. Adults average 20 to 30 millimeters in length, while the larvae are sometimes a bit larger. The pupa is a rapid swimmer but does not explode through the surface. Instead, it makes its way to partly submerged logs and rocks upon which it crawls out of the water and then begins its transformation. Needless to say, much commotion is created when trout, especially quality trout, take up pursuit. It is well worth the angler's efforts to do a little night fishing, or at least to rise very, very early to intercept the hatch.

During July and August when fishing is generally slow in the Midwest because of low, clear water conditions, several species of the genus *Nectopsyche* (formerly *Leptocella*), family Leptoceridae, present anglers in the Midwest an opportunity for the use of the old standard, the White Miller dry fly. This caddis is often called the White Miller in this area since it has white wings over its green abdomen. The pale green larva (6 to 10 millimeters) prefers areas of the stream where the flow of water is smooth and moderate. Adults are slim by comparison to other caddis flies and have excessively long antennae. They swarm in great numbers above pools and runs, appearing to fly backwards, an illusion created by their tail-like antennae.

Throughout the season many microcaddis, all members of the family Hydroptilidae, will be present. Only 2 to 5 millimeters in length, this group of caddis is not as widely distributed. Yet it is found in great numbers on spring streams where there is little fluctuation of water flow and temperature. Adults are very hairy and fuzzy; color is a flecked dark brown.

The most abundant of all caddis in the Midwest region, the net-building caddis of the family Hydropsychidae, emerge throughout the season. Hatching occurs almost daily from dawn to dusk. Soft-hackled wet flies are excellent imitations for the pupal form, even when no activity appears to be in progress. Adults average 10 millimeters in length and are mottled brown to dark brown. The larvae are active and exhibit very pronounced gilling on their abdomens. They build their nets in fast-flowing waters.

The Midwest also has good hatches of the largest of all caddis families, the Limnephilidae, in both running and still waters, intermittently throughout the season. Occasionally during the month of June there will be some very good hatches of ginger-colored limnephilids.

emergence table—midwest

	CLASSIFICATION	SEASON	TIME OF DAY
FAMILY	Philopotamidae	April/May	late morning/early afternoon
Genus	*Chimarra*		
FAMILY	Brachycentridae		
Genus	*Brachycentrus*		
species:	*numerosus*	April/May/early	late morning/
	americanus	June	afternoon
FAMILY	Leptoceridae		
Genus	*Nectopsyche*		
species:	several	May/September	sporadic late afternoon/ evening
		July/August	
Genus	*Mystacides*		
species:	*sepulchralis*	August/September	afternoon
FAMILY	Hydropsychidae		
Genus	*Hydropsyche*		
species:	*bifida*	throughout the	sporadic
	bronta	season	
	morosa		
species:	*simulans*	sporadic	sporadic
Genus	*Cheumatopsyche*		
species:	several	June/July	morning/afternoon
FAMILY	Lepidostomatidae		
Genus	*Lepidostoma*		
species:	*liba*	May/June	sporadic

ADULT INSECT SIZE (head to wing tip)	COLORATION			HOOK SIZE (ADULT)
7–10 mm	BODY: WING: LEGS:	black/white mottled dark grey dark bronze dun		#18–20
12–15 mm	BODY: WING: LEGS:	dark greyish olive mottled brown/grey dark bronze dun		#14–16
10–15 mm	BODY: WING: LEGS:	tannish grey mottled greyish tan dark ginger		#14–18
12–17 mm	BODY: WING: LEGS:	pale green whitish cream light ginger		#14–16
8–10 mm	BODY: WING: LEGS:	black metallic blue/black bronze dun		#18–20
9–12 mm	BODY: WING: LEGS:	greyish olive light brownish grey rusty dun		#16–20
13–15 mm	BODY: WING: LEGS:	brownish brown/grey mottled light brown		#14–16
7–8 mm	BODY: WING: LEGS:	bright green to olive medium grey rusty dun		#20–22
8–9 mm	BODY: WING: LEGS:	brown mottled tan/brown ginger		#20

continued

CADDIS COUSINS

emergence table—midwest *continued*

CLASSIFICATION	SEASON	TIME OF DAY
FAMILY Rhyacophilidae **Genus** *Rhyacophila* **species:** *lobifera*	May/June	sporadic
species: *ledra*	May/June	sporadic
FAMILY Glossosomatidae **Genus** *Glossosoma* **species:** *intermedum*	June/July	sporadic
FAMILY Hydroptilidae (microcaddis) **species:** several	throughout the season	sporadic
FAMILY Phryganeidae **Genus** *Ptilostomis*	June/July	early morning and night
FAMILY Limnephilidae **Genus** *Pycnopsyche*	August/September	evening/night
Genus *Neophylax* **species:** *concinnus*	June/September	evenings

ADULT INSECT SIZE (head to wing tip)	COLORATION			HOOK SIZE (ADULT)
11–13 mm	BODY:	greenish		#16–18
	WING:	brown/grey mottled		
	LEGS:	tannish		
8–10 mm	BODY	brownish		#18–20
	WING:	brown/grey mottled		
	LEGS:	light brown		
8–10 mm	BODY:	dark brown		#18–20
	WING:	greyish tan		
	LEGS:	rusty dun		
4–7 mm	BODY:	greyish to dark olive		#22–24
	WING:	grey brown		
	LEGS:	rusty dun		
21–24 mm	BODY:	rusty orange		#18
	WING:	rusty dark to light brown		
	LEGS:	ginger		
19–21 mm	BODY:	medium brown		#10–12
	WING:	mottled brown		
	LEGS:	mottled brown		
10–12 mm	BODY:	yellowish brown		#16–18
	WING:	mottled brown		
	LEGS:	dark rusty dun		

the west Though streams in both Idaho and Montana were researched directly by the authors, their time there was limited. Thus much of that research was done by friends, two in particular, Charles Brooks and Gary LaFontaine.

Charles Brooks, the nymph tier and angler who keeps taking larger and larger trout, fishes mostly nymphs—great, big stone fly nymphs. But he also knows a mountain about other flies, including caddis, and he is constantly re-searching.

Charlie writes that he does not carry identification of western caddis insects below the classification of genus due to the fact that nearly all species in a particular western genus have the same habits, habitat, and emergence characteristics. The variations are only in size, color, and emergence dates. He handles this situation by capturing both the immature and the emerging forms and tying his imitations to suit the occasion. His fly box carries various sizes and shadings of each genus so that when a particular species does come off, he can make his selection. He notes, "Emergence dates in my area *are not* relia-ble; we have a microclimate, and dates can differ by over a month for the same species at [nearly] the same spot." Even if you live in the West, and can keep current with the seasonal activity, I stongly suggest you take along a variety of sizes and colors to imitate the common caddis flies that emerge on these streams. Of course if you tie your own, you can simply do a streamside tie.

Charlie continues, "Nearly all our western caddis flies hatch by rising rapidly to the surface, bursting through and flying off. They come through the surface and are gone so quickly that it is almost impossible to make an identifi-cation. To give an idea of how difficult, it took me fifteen years to capture and identify one very tiny pattern on the Firehole. It hatches at dusk and returns to egg-laying during the night. It has caused me many sleepless nights because the trout were mad for it when it was hatching, but would not touch a floating fly."

To illustrate the complexities of western caddis hatches, here is Brooks' ac-count of the caddis activity on one stream, the Yellowstone in the Park, and other local rivers.

Brooks reports that the two most important western genera there are *Brachycentrus* and *Rhyacophila.* All species of *Brachycentrus* are the same color, but vary in size and, of course, emergence dates.

Brooks says that *Brachycentrus* begins to emerge on the Yellowstone in the Park on about July 20, two weeks either way, and continues until October, the early season hatches being larger.

"*Rhyacophila* hatch in the Yellowstone about two weeks later than *Brachycentrus* and also last until October. As the season progresses the two genera will overlap and the fish will take one in preference to the other. There is no way of knowing which one and thus experimentation is in order. Flying caddis appear to be larger than they actually are, and lighter in color. We try to spot and identify them when they are on the water. Egg laying time on the Yellowstone is generally from mid-afternoon until dark.

"The Yellowstone has a hatch of about a #10 [limnephilid], species unknown. The hatch is scattered and irregular but the imitation is very effective at egg laying time (until 10 in the morning and after dark).

"The Firehole has the same species of *Brachycentrus* and *Rhyacophila* as the Yellowstone but they hatch in May, June and July. They are scattered along the river . . . long stretches do not host either genus. Sizes are generally 14, 16, 18. The *Chimarra* genus is abundant on the Firehole and hatches in August and September.

"On the Madison in the park *Brachycentrus* and *Rhyacophila* are the general rule by August 1. Sizes are the same as the Yellowstone except that the *Brachycentrus* does not get down to a size 18. An early season imitation is good for either genus in this section of the river."

Henry's Fork: "No one knows how many genera and species of caddis inhabit this wonderful river. I believe it contains more species than all the foregoing mentioned rivers combined. *Brachycentrus* and *Rhyacophila* are the most important, but far from the only ones. At Railroad Ranch, these two genera hatch during June, July and August . . . and perhaps as late as September 15 in warm areas.

"*Brachycentrus* and *Rhyacophila* are abundant in the lower Madison, from Quake Lake to Ennis. They are difficult to fish because of the type of water, but effective along the edges and quieter areas about dusk. The same holds true for the Gallatin in Yellowstone Park."

So you can see that the Western fly fisherman has to be a pioneer to fish the caddis hatches successfully. Nonetheless, the emergence charts that follow will outline the possibilities for hatches in the mountain streams, and give a key to caddis patterns that are likely to be effective.

emergence table—west

	CLASSIFICATION	SEASON	TIME OF DAY
FAMILY Genus	Goeridae *Goera*	late May/July	morning/evening
FAMILY Genus species:	Brachycentridae *Brachycentrus* *americanus*	late June/July	morning/afternoon
species:	*occidentalis*	late June/July	morning/evening
FAMILY Genus species:	Hydropsychidae *Hydropsyche* *californica* *occidentalis* *oslari*	late July/ September	morning/afternoon
Genus species:	*Arctopsyche* *grandis*	July/August	sporadic
Genus · species:	*Cheumatopsyche* *gracilis*	late June/July	afternoon/evening
FAMILY Genus species:	Lepidostomatidae *Lepidostoma* *pluviale*	July/August	sporadic
FAMILY species:	Hydroptilidae (microcaddis) several	June/October	sporadic
FAMILY Genus species:	Leptoceridae *Oecetis* *disjuncta*	June/September	afternoon/evening

ADULT INSECT SIZE (head to wing tip)	COLORATION		HOOK SIZE (ADULT)
14–16 mm	BODY:	brownish yellow	#14–16
	WING:	mottled grey	
	LEGS:	mottled grey	
13–15 mm	BODY:	greyish olive to brownish grey	#14–16
	WING:	brownish grey	
	LEGS:	bronze dun	
11–13 mm	BODY:	brownish grey	#16–18
	WING:	grey brown mottled	
	LEGS:	rusty dun	
9–11 mm	BODY:	light brownish	#18–20
	WING:	dark mottled tan	
	LEGS:	ginger	
10–12 mm	BODY:	greyish olive	#16–18
	WING:	light brownish grey	
	LEGS:	rusty dun	
12–14 mm	BODY:	dark brown	#14–16
	WING:	brownish grey	
	LEGS:	bronze dun	
12–14 mm	BODY:	dark tannish	#16
	WING:	mottled grey	
	LEGS:	rusty dun	
8–9 mm	BODY:	slate grey	#20
	WING:	dark grey brown	
	LEGS:	grey	
3–6 mm	BODY:	olive grey	#22–24
	WING:	mottled grey	
	LEGS:	rusty grey	
10–11 mm	BODY:	dark reddish brown	#18
	WING:	slate grey	
	LEGS:	brownish dun	

continued

emergence table—west *continued*

CLASSIFICATION		SEASON	TIME OF DAY
FAMILY	Rhyacophilidae		
Genus	*Rhyacophila*		
species:	*acropodes*	July/October	morning/afternoon
species:	*verrula*	August/September	sporadic
FAMILY	Glossosomatidae		
Genus	*Glossosoma*		
species:	*montana*	late July/	sporadic
	excita	August	
	alascense		
FAMILY	Helicopsychidae		
Genus	*Helicopsyche*		
species:	*borealis*	late June/	afternoon/evening
		September	
FAMILY	Polycentropodae		
Genus	*Polycentropus*		
species:	*cinereus*	June/September	late morning/
			afternoon
FAMILY	Phryganeidae		
Genus	*Phryganea*		
species:	*cinerea*	August	evening
	(high lakes and slow		
	river pools)		
FAMILY ·	Limnephilidae		
Genus	*Phychoglypha*		
species:	*ulla*	September/	late afternoon/
		October	evening
Genus	*Neophylax*		
species:	*splendens*	September/	late afternoon
		October	
Genus	*Limnephilus*		
species:	*externus*	September/	evening
	(high elevation lakes	October	
	and pools)		
Genus	*Dicosmoecus*	late September/	late evening/night
		October	

ADULT INSECT SIZE (head to wing tip)	COLORATION		HOOK SIZE (ADULT)
11–12 mm	BODY: WING: LEGS:	greenish brown/light grey mottled rusty dun	#16–18
13–14 mm	BODY: WING: LEGS:	dark tan mottled ginger ginger	#16
8–10 mm	BODY: WING: LEGS:	dark brown greyish tan brown grey	#18–20
5–7 mm	BODY: WING: LEGS:	tannish grey brownish grey rusty dun	#22
8–9 mm	BODY: WING: LEGS:	brownish tan mottled brown tan ginger	#20
20–25 mm	BODY: WING: LEGS:	olive grey greyish brown rusty dun	#8–10
15–17 mm	BODY: WING: LEGS:	light brown light brown with silvery strip ginger through center of front wing	#14
14–15 mm	BODY: WING: LEGS:	greyish brown mottled grey brown tannish grey	#14–16
19–21 mm	BODY: WING: LEGS:	brownish olive mottled brownish rusty dun	10–12
29–35 mm	BODY: WING: LEGS:	orange brown greyish brown rusty dun	#6–8

far west   Dave McNeese in Vaneta, Oregon.

Dicosmeocus . . . Oregon . . . Northwest . . . October . . . big trout!

When I think of Dave McNeese my mind automatically assembles to the streams in Oregon and Washington and the huge caddis fly *Dicosmeocus,* of the family Limnephilidae; I picture large trout chasing the largest and slowest of all our caddis flies. Lumbering like a heavily-laden moth on the surface of the stream, the emergence of this genus provokes and agitates the most cautious of trophy trout. Commonly called the orange sedge, *Dicosmeocus* emerges during the afternoon with the hatch continuing until dark. This activity continues for approximately three weeks. It brings out not only the trout, but more anglers as well.

Dave McNeese has done a mountain of research on far western caddis. He has collected samples of caddis from various streams (cases and adults), from the time the season opened until it closed (late April to October). In addition to this time-consuming undertaking he has sent numerous pages of notes regarding general information of caddis activity in the Northwest in addition to flies and pattern listings (see Chapter 7) of imitations that will be useful to the angler. Here are some of his findings:

Most rivers in the western slope of the Northwest region are generally high and milky during the last week in April, while the eastern portions of the range are fairly clear, or clearing. Very little caddis activity takes place during this time. Samples taken from the stomachs of trout show a high percentage of mayfly and stonefly larvae, but very few pupae.

Throughout May as the water temperature gradually increases a few more caddis are found. A good imitation at this time is a size 14 brown caddis fished at dusk, especially in a pupal imitation. Dave advocates a weighted pupa cast across and downstream, and twitched upstream on the swing until the fly is aligned with the tip of the rod.

Late May and early June produce a good hatch of large green caddis, *Brachycentrus,* at the lower elevations. The McKenzie Special (bright green floss body, brown bucktail wings and a grizzly hackle) is a favored imitation. The green caddis, because of its body weight, is a clumsy flier, thus highly attractive to trout, especially large ones. Emergence takes place in late afternoon until dark (size approximately 18 mm).

Following the green caddis is the hatch of the cinnamon caddis. This is a very fast fly; best fishing occurs at dusk during ovipositing. Dave prefers to use a long rod with which he can control the fly while he bounces it off the surface in shallow riffles (two to three feet deep), similar to the fluttering action described in Chapter 4). The cinnamon caddis generally lasts through June, though at higher elevations is active into July.

June is the most active month of the season for the fly fisher, since not only are numerous caddis flies present, but mayflies and stoneflies are at their height of emergence. With complex hatches continually developing it becomes a bit more difficult for the angler to determine just exactly what trout are feeding on.

During the last week in June a "superhatch" (probably *Hydropsyche*) of a small green caddis takes place, lasting through July. It is complicated since it coincides with a large stonefly hatch of *Isoperla,* which many anglers prefer to imitate. This is quite natural, since common sense leads us to believe that trout will feed on the largest available insect present. However, Dave has had more success by fishing a size sixteen green caddis using a short line. Both adult and pupal imitations are used. When fishing the pupa he recommends using a very short line and fishing the imitation as an emerger in the manner of the Leisenring lift, or down-and-across stream cast.

By the time the small green caddis has disappeared in mid-August a fair hatch of brown caddis emerges. This at a time when most of the mayflies have had their day in the sun. This especially holds true in streams wherein the temperature has gone above 58°. The brown caddis is tied in a size 14 to 18 with success into September.

The season ends with a bang with the appearance of the orange sedge *(Dicosmeocus)* in October. When this hatch begins trout will feed exclusively upon this insect.

emergence table—far west

CLASSIFICATION		SEASON	TIME OF DAY
FAMILY Genus species:	Rhyacophilidae *Rhyacophila* *acropodes*	late May/July	late morning/ afternoon
species:	*verrula*	August/September	sporadic
FAMILY Genus species:	Glossosomatidae *Glossosoma* *excita* *alascense*	late June/July	sporadic
FAMILY Genus species:	Brachycentridae *Brachycentrus* *americanus*	late June/July	morning/afternoon
species:	*occidentalis*	June/July	morning/afternoon
FAMILY Genus species:	Hydropsychidae *Hydropsyche* *californica* *occidentalis* *oslari*	July/September	morning/afternoon or evening
Genus species:	*Arctopsyche* *grandis*	late June/July	morning/afternoon
FAMILY Genus species:	Lepidostomatidae *Lepidostoma* *pluviale*	July/August	sporadic

ADULT INSECT SIZE (head to wing tip)	COLORATION		HOOK SIZE (ADULT)
11–12 mm	BODY: WING: LEGS:	greenish brown/light grey mottled rusty dun	#16–18
13–14 mm	BODY: WING: LEGS:	dark tan mottled tan ginger	#16
8–10 mm	BODY: WING: LEGS:	dark brown greyish tan brown grey	#18–20
13–15 mm	BODY: WING: LEGS:	greyish olive brownish grey brown dun	#14–16
11–13 mm	BODY: WING: LEGS:	brownish grey grey brown mottled rusty dun	#16–18
9–13 mm	BODY: WING: LEGS:	light brownish dark mottled tan ginger	#18–20
11–13 mm	BODY: WING: LEGS:	greyish olive light brownish grey rusty dun	#16–18
12–14 mm	BODY: WING: LEGS:	dark brown brownish grey bronze dun	#14–16
8–9 mm	BODY: WING: LEGS:	slate grey dark grey brown greyish	#20

continued

CADDIS COUSINS

emergence table— far west *continued*

	CLASSIFICATION	SEASON	TIME OF DAY
FAMILY	Hydroptilidae (microcaddis)	July/September	sporadic
FAMILY Genus species:	Leptoceridae *Oecetis* *disjuncta*	July/September	afternoon/evening
FAMILY Genus	Limnephilidae *Oligophlebodes*	July/August	evening
Genus species:	*Neophylax* *splendens*	July/August	late afternoon/ evening
Genus	*Dicosmoecus*	late September/ October	late evening/night

ADULT INSECT SIZE (head to wing tip)	COLORATION		HOOK SIZE (ADULT)
4–7 mm	BODY:	olive grey	#22–24
	WING:	brown grey	
	LEGS:	rusty dun	
10–11 mm	BODY:	dark reddish brown	#18
	WING:	slate grey	
	LEGS:	bronze	
10–11 mm	BODY:	yellowish brown	#18
	WING:	light brown mottled	
	LEGS:	tan	
14–15 mm	BODY:	greyish brown	#14–16
	WING:	mottled grey brown	
	LEGS:	tannish grey	
29–35 mm	BODY:	orange brown	#6–8
	WING:	greyish brown	
	LEGS:	rusty dun	

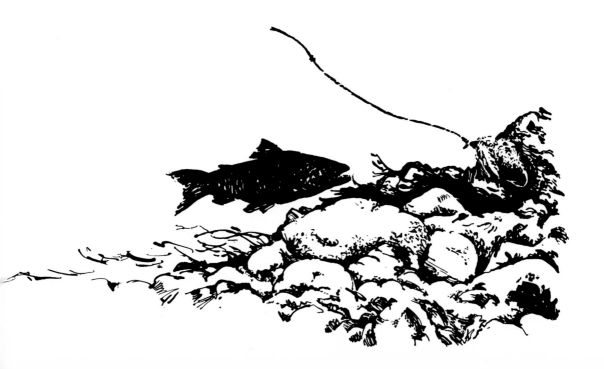

relationship of antennae length to length of insect

(from head to wing tip)

FAMILY	GENUS	shorter	same	longer
Brachycentridae	*Brachycentrus*	X		
Goeridae	*Goera*	X		
Helicopsychidae	*Helicopsyche*	X		
Hydropsychidae	*Hydropsyche*		X	
	Cheumatopsyche		X	
	Macronema			X
	Arctopsyche		X	
Lepidostomatidae	*Lepidostoma*	X		
Leptoceridae	*Nectopsyche*			X
	Ceraclea			X
	Oecetis			X
Limnephilidae	*Pycnopsyche*	X		
	Limnephilus	X		
	Hesperophylax	X		
	Neophylax	X		
	Dicosmoecus	X		
	Astenophylax	X		
	Oligophlebodes	X		
Odontoceridae	*Psilotreta*		X	
Philopotamidae	*Chimarra*	X		
	Dolophilodes	X		
Phryganeidae	*Phryganea*	X		
	Ptilostomis	X		
Polycentropodidae	*Polycentropus*	X		
Rhyacophilidae	*Rhyacophila*	X		
	Glossosoma	X		

One of the two most visible identification marks for distinguishing caddis families is the shape of the adult wing. The outlines here may help you make a preliminary identification of adult naturals, so that you can make important judgments about fishing the ovipositing females (see Chapter 5), and the larva as well (Chapter 2). The antennae length chart will, in some cases, allow you to take the wing shape group one step farther, perhaps to a definite family and genus. Caddis adults are extremely hard to identify by species, and for the angler's purposes, all that is important is to determine whether the particular insect is free-living as a larva, and how it will behave during the ovipositing stage. These charts represent a shorter way of making this identification, for some families.

caddis wing silhouettes

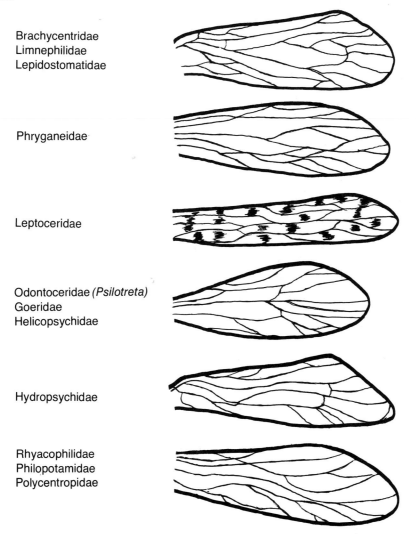

Brachycentridae
Limnephilidae
Lepidostomatidae

Phryganeidae

Leptoceridae

Odontoceridae *(Psilotreta)*
Goeridae
Helicopsychidae

Hydropsychidae

Rhyacophilidae
Philopotamidae
Polycentropidae

some typical adult caddis

Drawings of typical caddis family members taken from *The Caddis Flies, or Trichoptera, of Illinois* by Herbert H. Ross. Ross' study is unsurpassed as a primary source of information about caddis from an entomological point of view. *(Used by permission of The Illinois Natural History Survey.)*

Drawings by C. O. Mohr

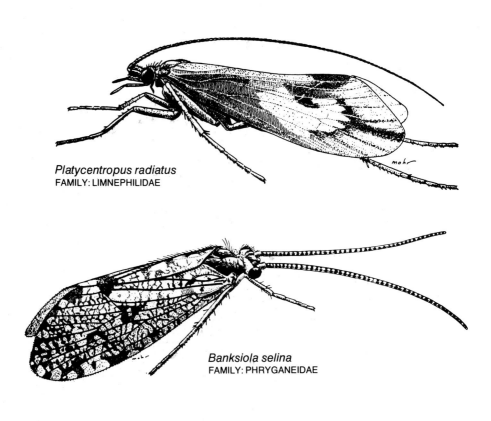

Platycentropus radiatus
FAMILY: LIMNEPHILIDAE

Banksiola selina
FAMILY: PHRYGANEIDAE

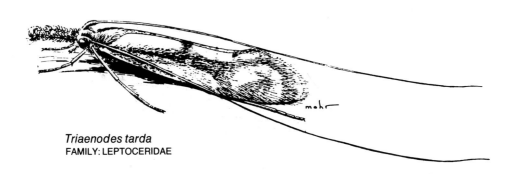

Triaenodes tarda
FAMILY: LEPTOCERIDAE

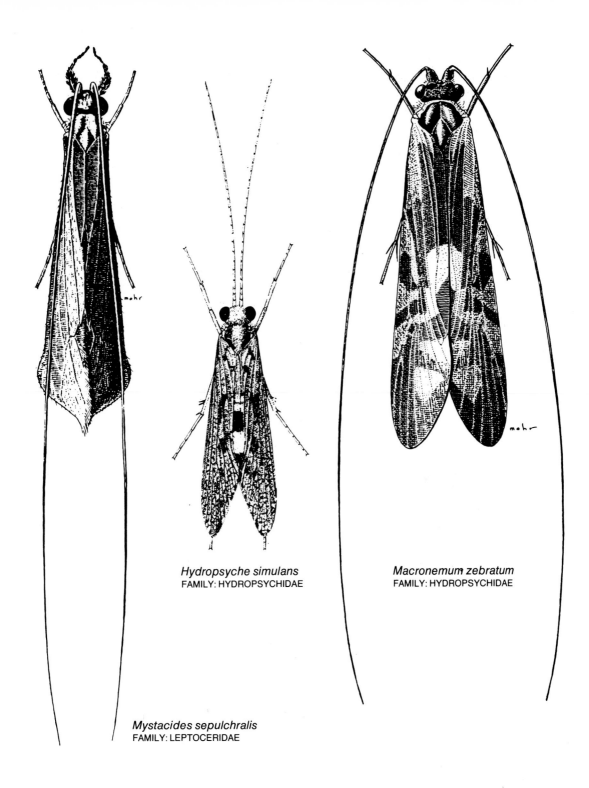

Hydropsyche simulans
FAMILY: HYDROPSYCHIDAE

Macronemum zebratum
FAMILY: HYDROPSYCHIDAE

Mystacides sepulchralis
FAMILY: LEPTOCERIDAE

CADDIS COUSINS

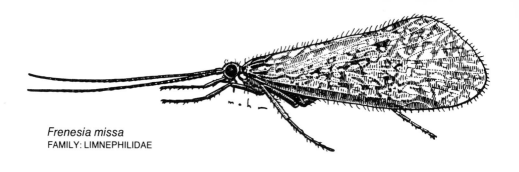

Frenesia missa
FAMILY: LIMNEPHILIDAE

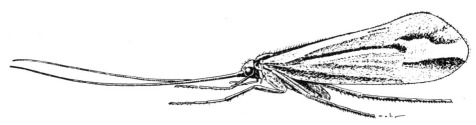

Hesperophylax designatus
FAMILY: LIMNEPHILIDAE

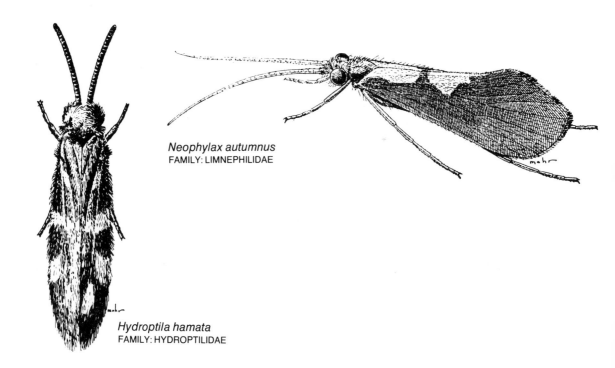

Neophylax autumnus
FAMILY: LIMNEPHILIDAE

Hydroptila hamata
FAMILY: HYDROPTILIDAE

favorite local patterns

What is an expert? The dictionary states that an expert is a person who has a special skill or knowledge regarding a specific subject; one who has mastered his craft to a certain high degree of proficiency, or has knowledge far beyond the average in his particular field.

We have all heard of fishing experts, and I suppose there are such individuals. I don't pretend to be one of them, nor do I want to become one. If the day ever comes that I can consistently outwit trout, or any other fish, I will look to other fields for another challenge. The one sure thing about fishing is that it is not a sure thing—and that's why it's fun.

One thing I do know about our sport, however, is that I can always learn from other anglers in our fraternity. And that's what this chapter is primarily about. For ironically, even though caddis have never, until recently, gotten their real share of attention in the fly fishing tradition, over the years anglers have gone about experimenting with new ties on a local level. And since most caddis emergences are a local affair with local names (or none at all), many of these very worthwhile patterns have never been given wide distribution. Nonetheless, these caddis patterns have been developed in the best spirit of fly fishing innovation, and are most often the result of personal experimentation, even though their originators were not known to be "experts."

A great deal of time was spent contacting anglers throughout the states to get their proven caddis patterns. And, it was worthwhile. They are *my* experts!

The only condition exacted of each angler who contributed information to use was that the patterns sent be those that he found truly effective; that they represented caddis imitations that worked in his area. Since caddis patterns like these are necessarily the products of individual observation and experience, testing their value to fly fishermen in other parts of the country could be a fascinating process in itself. And, the subsequent list should keep the angler tier fairly busy.

Anglers in Colorado probably fish more caddis imitations than mayfly patterns. In Colorado the caddis is king. In fact, one of the most popular caddis flies of the state, not counting the Henryville Special, is named the Colorado King—not only because GEORGE BODMER of Colorado Springs, the originator of this pattern, says so, but the fact that it is acclaimed as such by most of his competitors and friends.

Prior to speaking to George one afternoon I called KEN WALTERS of Denver and NORMAN LEIGHTY of Boulder. The first fly they mentioned as a local favorite was the Colorado King. In addition to the King, Ken Walters guessed that the PM Caddis was also a very good producer while Norm Leighty sang the praises of two old standbys, the Henryville Special and the Gold-Ribbed Hare's Ear—the latter, of course, used as a pupal imitation. On the other side of the range in Aspen, Colorado, CHUCK FOTHERGILL also touted the Gold-Ribbed Hare's ear as his all time favorite.

colorado king (light): *(tied in sizes 10 to 18)*

hook:	Mustad 94840 or 7957B
thread:	black, 6/0 pre-waxed
tail:	two peccary fibers, tied long and spread at wide angle
body:	yellow rabbit, or Fly Rite #9 or #14
rib:	grizzly hackle palmered through body (6–7 turns)
wing:	elk hair (or deer) tied downwing style, slightly longer than shank
head:	black, varnished

The Colorado King is tied in a number of color versions. The following three are equally as popular.

dark colorado king

Tied in the same manner, except that the body is made from muskrat fur (or Fly Rite #8). The wing should be a slightly darker natural tan brown piece of elk hair.

brown colorado king

Same as the Colorado King, except body is medium brown dubbing from a Hare's Mask (or Fly Rite #20). The rib of hackle is brown and the wing should have a brownish cast.

egg sac colorado king

Same as the dark version except that an egg sac of yellow rabbit (or Fly Rite #9) is tied in at the bend.

The Colorado King was developed in 1971 and has produced not only in the Rockies, but throughout the world. It was designed as an all-purpose caddis and has the functional characteristics of having a good floatation, high visibility, durability, and ease of tying. George tells me that he knows the natural caddis fly does not have a tail. However, the purpose of the peccary tail in his pattern is for stability. It acts in the manner of an outrigger on a canoe, keeping the fly high, and on even keel.

Ken Walters described the PM Caddis to me as follows:

pm caddis *(tied in sizes 10 to 18)*

hook:	Mustad 94840
body:	olive poly dubbing
wing:	natural brown deer hair (fine)
hackle:	grizzly and brown mixed

The PM Caddis is also tied with a dark grey body.

If you are ever in that area, Norm Leighty recommends trying the patterns mentioned on the South Platte River near the town of Dexter, should you desire some very enjoyable fishing.

GARY BORGER lives in Wausau, Wisconsin, but he fishes all over the state, and all over the country. Like so many contributors to this volume, he is also a good friend.

Two years ago Gary sent me a caddis imitation called "The Eddy Fly", or "Eddie's Fly". I'm not sure which name is correct, nor does Gary remember; the only thing we do know is that it serves as an all-purpose caddis imitation. In fact, I have had one of my finest days of fishing using this pattern. It is very simple to tie. It is very dry for its seeming sparseness of body.

eddy fly *(tied in sizes 10 to 14)*

hook:	Mustad 94840
tip:	yellow monocord
body:	burnt orange monocord
wing:	woodchuck guard hairs
hackle:	brown and grizzly mixed

Because of the limitation of color barring on woodchuck guard hairs, this fly cannot be tied in smaller sizes using this material. However, I have found that the barring on the tail of a common chipmunk is identical to that of woodchuck guard hair, but in much smaller segments. Thus, if you substitute chipmunk tail hair fibers, this pattern can be tied down to sizes 16 and 18.

One would not think a fly with only a body of thread and a wing of woodchuck would float high, but it does. It is the woodchuck fibers that add to its floatation abilities.

"Why don't you come up and fish with me. I think I can guarantee you some steelhead."

"I'd like to, Larry," I said, "but things are just a bit hectic this year." (Why is it that things are always hectic *this* year?)

LARRY BROOKS lives in Bellingham, Washington, just below the Canadian border. The conversation went on. The fishing got better . . . and I felt worse.

"Listen, Larry," I cut in, "you're just making my life a little tougher here in the East, so don't tell me how great it is out there. While I've got you, I need a favor. I want to know the most productive caddis patterns from your area. The kind you use for trout."

"Trout? . . . oh, trout!" he was excited. "Gee, I wish there were more trout to fish for; all we ever get is steelhead."

The poor soul. I felt so sorry for him I would have traded places instantly just so he could have a change of scenery.

In any event, the caddis patterns arrived: two pupae. His favorite is the Dandy Green Nymph.

dandy green nymph *(tied in sizes 10 to 16)*

hook:	Mustad 3906B
tail:	dull light blue hackle fibers or pheasant body
body:	dull blue monofilament ribbing
hackle:	natural guinea fowl or grouse neck feathers
head:	lacquered white

rainbow royal *(tied in sizes 10 to 16)*

hook:	Mustad 3906B
tail:	dull blue pheasant body (lower portion) fibers
body:	bright green wool or fur
rib:	fine gold oval tinsel
hackle:	grouse (mottled) or guinea hen natural

ALLEN BREAULT, of Fitchburg, Massachusetts went to great lengths in designing a caddis pupa imitation which he feels is superior to the conventional type in which the wings are made from duck wing quill section. The wings on Allen's pattern are made from rooster hackles on which the barbules have been forced rearward, thus creating a veined effect. (This procedure has been used for some time on dry flies like the Miracle Wing.)

The fly certainly does have a realistic appearance. Allen has fished it with great success on the Beaverkill, the Battenkill, and the AuSable, three of the hardest-fished streams in New York. Here is its description. This is a general fly pattern. With the exception of the head, colors are not fixed. Colors for the body, legs, and wings should be chosen to represent the natural being imitated.

vein winged caddis pupa *(tied in sizes 8 to 16)*

hook:	for large flies Mustad 38941 or 79580; for size ten and smaller, Mustad 3906 or 94840
underbody:	prewaxed dental floss (for bulk)
body:	water animal dubbing fur
rib:	Maxima monofilament
legs:	mallard flank
wing:	rooster hackle with fibers tied rearward
head:	mini ostrich (dark brown or black)

ANDREW CAMBO, of Winsted, Connecticut, writes, "These will catch trout anytime; when nothing seems to work I tie on. . . ."

Four caddis patterns accompanied the letter. Frankly, I'm a believer that there is no fly that works all the time. I also believe there are times that all the flies ever originated will not work during those rare (thank God) days when trout just will not co-operate.

Nevertheless, such confidence in a few patterns deserves mention; here is the listing.

olive/brown caddis *(tied in sizes 12 to 16)*

hook:	Mustad 94840
body:	olive (medium to dark) fur
rib:	cream hackle palmered and trimmed next to body (stubby effect)
wing:	mottled brown grouse or pheasant quill, flat and tip "V" notched
hackle:	brown

light olive/grey caddis *(tied in sizes 12 to 18)*

hook:	Mustad 94840
body:	light olive wool or fur
wing:	slate grey duck quill tied flay and V-notched
hackle:	brown

olive/grey delta wing *(tied in sizes 12 to 18)*

hook:	Mustad 94840
body:	olive (light)
wing:	light dun hackle tips tied Delta style
hackle:	brown

brown caddis: *(tied in sizes 12 to 18)*

hook:	Mustad 94840
body:	brown floss
rib:	cream hackle palmered and trimmed to stubs
wing:	mottled brown grouse or pheasant quill tied flat and V-notched
hackle:	brown

TOM DARLING of Seattle, Washington, spends most of his life in animated excitement when he is with friends of our angling fraternity (or lost in travels) rushing from one stream to another to see if the water isn't just a bit clearer. When I asked him for some caddis patterns, his enthusiasm almost overwhelmed me and I had to do some fast talking to get him down from twenty-seven patterns to a few select handful. After much decision-making, I also got him to name his favorite; it is listed first.

vincent sedge *(tied in sizes 10, 12)*

tail:	deer hair
body:	peacock colored polypropylene
underwing:	deer hair
wing:	folded brown turkey or duck quill that has been sprayed with Tuf-Film (trimmed to V shape)
hackle:	brown or cree (use two)

deschutes caddis *(tied in sizes 8 to 12)*

hook:	Mustad 94840
tail:	deer hair tied short
body:	medium yellow wool
wing:	light brown deer hair
hackle:	dark ginger

grey sedge *(tied in sizes 8 to 10)*

hook:	Mustad 94840 or equivalent
tail:	red hackle fibers
body:	grey wool or fur
rib:	fine oval silver tinsel
wing:	mallard flank feathers, tied flat
hackle:	dark badger

light caddis *(tied in sizes 8 to 16)*

hook:	Mustad 94840 or equivalent
tail:	ginger hackle fibers, tied short
body:	cream wool
rib:	golden ginger hackle palmered
wing:	natural grey duck quills, tied slightly Delta Wing.

note: Many of the Northwest caddis patterns listed seem to call for a tail on the imitation. However, the wing on most of these flies is tied no longer than to the bend. Thus, the silhouette present to the trout is very close to that of a natural caddis on which the wing is half again as long as the body. Caddis have no tail, and the "tail" on these flies actually simulates the wing extension.

A very simple pattern to tie, and yet one of the all time favorites on New York's Beaverkill is WALT DETTE's Conover. Here is his pattern description.

conover *(tied in sizes 10 to 16)*

hook:	Mustad 94840
tail:	badger
body:	muskrat and claret
hackle:	badger

I first met JACK GARTSIDE in West Yellowstone. He was camped in Yellowstone Park by the banks of Duck Creek where he had been roughing it along with his cat Merlin, in a small trailer which was crammed with all his possessions. I don't know which place he calls home. Formerly a Boston school teacher, he now locates wherever his whims take him. The last time I saw Jack was in Sun Valley, Idaho, and his destination from there was British Columbia . . . for the time being.

Jack is a student of nature; his directions for patterns are taken from the insects themselves, the ones trout feed upon in their particular region. The Pheasant Caddis is one he originated for Montana fishing.

pheasant caddis

hook:	Mustad 94840
thread:	black
body:	tan dubbing
rib:	palmered dark ginger hackle clipped to hook-gap size
wing:	mottled body feather from hen or cock pheasant (light shade), which has been lacquered. This is of one piece with a V-notch cut into the rear portion.
hackle:	brown (or dark ginger)

Jack varies the sizes and colors according to the species of caddis emerging. For a darker imitation the darker pheasant body feathers (mottled), are employed. Hackle is then changed to a grizzly and brown mix. Body colors can be cream, olive, or grey in addition to the brown or tan. Favorite furs used are otter, beaver, rabbit, or "whatever".

JOHN HARDER, who lives in Manchester, Vermont, has fished many of New England's waters. In addition to helping us collect samples of caddis larvae, he has sent us a list of five of his favorite patterns.

brown and grizzly bivisible
(tied in sizes 14 to 20)

hook:	94838 (Mustad)
body:	rear two thirds, grizzly; remaining third to head, golden ginger hackle

I suppose this pattern proves how very simple a fly can be and still produce. The two types of hackle are simply tied to a bare shank. This pattern and the other four flies contributed by John have all been tested and proven.

adams fluttering caddis *(tied in sizes 14 to 18)*

hook:	94840 (Mustad)
body:	muskrat belly
wing:	medium blue dun mink tail guard hairs
hackle:	brown and grizzly mixed

The conventional Adams has long been used during a caddis hatch; what makes this particular tie different is the use of the mink tail guard hairs for the wing as opposed to the fragile hackle tips.

brachycentrus emerger *(tied in sizes 12 and 14)*

hook:	Mustad 3906B
body:	natural brown seal fur tied full
rib:	size #36 copper wire
hackle:	brown partridge, tied as a collar

This pattern is fished as an emerging pupa. It can, of course, be changed in color to imitate other species.

hermaphrodite emerger *(tied in sizes 12 to 16)*

hook:	Mustad 3906
body:	natural brown weasel with guard hairs mixed in
wing:	mallard primary, tied tent-shaped
hackle:	golden ginger

John asked me to be sure to mention the fact that the Hermaphrodite pattern was credited to Al Brewster of Riverside, Rhode Island.

smokey blue dun fluttering caddis
(tied in sizes 14 to 18)

hook:	Mustad 94840
body:	pale beige mink fur
wing:	beige mink tail guard hairs
hackle:	beige or smokey dun, or dirty white

ELLIS HATCH lives in Rochester, New Hampshire. We've talked over the phone many times and also corresponded by mail; I have a standing invitation to fish with him, anytime. We've become close friends, and yet, we've never met. (I promised him that as soon as this work was finished we would wet a line together. I will keep that promise.)

When I called him recently concerning the caddis activity in his area I was informed that a caddis imitation was one of his favorite types of dry fly fishing. When I asked him for a pattern description, he said, "It doesn't have a name, but it sure works.

hatch "no-name" caddis: *(tied in sizes 12 to 18)*

hook:	Mustad 94840
body:	black seal fur
wing:	black deer hair
hackle:	none

This pattern is also tied using natural cream seal fur with a white deer hair wing.

During the course of the conversation Ellis mentioned the use of the ever popular Hornberg pattern as an imitation of a large caddis fly that emerges during early summer. Actually the Hornberg closely resembles the adult caddis species in the family Phryganeidae.

Another popular pattern used locally was the Gold-Ribbed Hare's Ear, which is fished to imitate an emerging pupa.

TERRY HELLEKSON and BRUCE BARKER, both of Ogden, Utah, sent not only a sample (excellently tied) of a larva, but also a bottled specimen of the natural they were imitating with this pattern. The pattern is called Free Living Caddis. Its description reads thus:

free living caddis *(tied in sizes 8 to 14)*

hook:	Mustad 79580
thread:	brown
tail:	small tuft of tan maribou (not past bend)
body:	tan synthetic fur over which a weaving process is employed using tan nylon hair for the back and clear round monofilament for the abdomen
thorax:	dubbed tan synthetic fur over which a wing case of moose hair forms the top
legs:	three moose body hairs tied in at each side and clipped to proper length
head:	same as thorax

note: The snythetic fur on the body is picked out and then clipped flat on the bottom. It is also trimmed at the sides to simulate gills.

I suppose the variations of the old Henryville Special are endless. There always seem to be improvements and versions that work better than the original. But there is also an old angling saw which goes, "you take more fish with the flies you like because *you fish them with more confidence*." I'll agree with that.

JAMES HEPNER of Sunbury, Pennsylvania, probably also agrees with it. This Henryville has got them all beat, he says.

jim's henryville #2 *(tied in sizes 10 to 24)*

hook:	Mustad 94840 or equivalent
body:	fiery brown (fur or poly dubbing)
rib:	palmered grizzly
wing:	grizzly hackle tips tied in open V-style (this is very close to Delta Wing style)
hackle:	brown

Jim says to get the body well down into the bend. It is tied in body colors of cream, olive, black and grey in addition to the above. The grizzly hackle tip Delta Wings are also dyed into shades of olive or brown depending on the natural.

WARREN JOHNS, of Portsmouth, Rhode Island, ties a unique imitation he calls the Caddis Tail. Prior to the development of this unique fly, Warren would wing his patterns with sections of mottled turkey wing quills, or duck primary feathers. Sections were cut from these feathers, lacquered, and a V-shaped notch cut into the tip in order to imitate the natural. Even though heavily lacquered, most of the flies did not last very long after a trout took them; the wing always split. I suppose Warren felt he was not getting his money's worth from the patterns he tied. In any case, with his new design, he not only gets longer wear from each fly he ties (a record of thirty-three trout on one fly) but also obtains better results.

Warren used the same material. However, instead of tying in a long section as a wing that extends beyond the bend, Warren ties in a very short section *at the bend,* as you would a tail. The rest of the fly is finished in the normal manner. If you were to look at one of Warren's flies while simultaneously viewing a conventional fly of the same type from underneath, you would see no difference between the two. And the trout sees them the same way. But the Caddis Tail gets additional effects from the broken light particles filtering through the wingless pattern, thus giving the imitation an impression of life.

Here is Warren Johns' pattern; various color combinations should be used.

caddis tail fly *(tied in sizes 12 to 18)*

hook:	Mustad 94840
tail:	V-cut section of lacquered mottled turkey
body:	greyish brown
rib:	grizzly hackle open palmer
hackle:	brown

FRANK JOHNSON of Missoula, Montana, has a sleepy smile and an easy-going personality that will make you like him the first time you meet. He appears to be able to handle a twenty-inch brown with nonchalance, while I can't handle fourteen inches without uncontrolled excitement.

When I asked him for a couple of good Montana caddis patterns he came up with two of his favorites.

bucktail caddis *(tied in sizes 10 to 16)*

hook:	Mustad 94840 or equivalent
tail:	short brown deer hair
body:	tan poly yarn, or poly dubbing
rib:	brown hackle palmered
wing:	deer hair, brown natural
hackle:	brown

goddard caddis *(tied in sizes 10 to 16)*

hook:	Mustad 94840 or equivalent
body:	deer body hair, spun and trimmed to caddis shape
wing:	deer hair, dyed dark tan
antennae:	stripped hackle stems from brown hackle feather
hackle:	brown

Listed below are two of the patterns GARY LA FONTAINE has devised incorporating the use of sparkle yarn. Both patterns are used during the emergence of the genus *Brachycentrus*.

grannom pupa *(tied in sizes 10 to 14)*

hook: Mustad 94840
overbody: chocolate brown sparkle yarn
underbody: blend of dark speckle of hare's ear and chocolate brown sparkle
hackle: dark grouse fibers
head: dark brown marabou strands
thread: brown

emergent grannom *(tied in sizes 12, 14)*

hook: Mustad 94840
overbody: chocolate brown sparkle yarn partially frayed
underbody: olive/brown rabbit fur blend
wing: brown speckled tips of deer body hair spread over back
head: dark brown marabou strands
thread: black

NAT LONG, of Amawalk, New York, was at the time of this writing, seventy-six years of age. His features, however, belie his age, which he attributes to the fact that he fishes or hunts every day except Sunday which is, of course, a day of rest.

Nat has fished and hunted for nearly sixty years. And for most of those sixty years he has unselfishly shared his knowledge with others. I don't know of anyone, except George Harvey, the noted Pennsylvanian, who has taught more people, young and old alike, to fish and tie flies.

A number of flies have been named in honor of Nat Long, among them a caddis imitation that he had used long before most of us knew there was anything other than a mayfly in our streams.

nat's caddis *(tied in sizes 12, 14, and 16)*

body: cream wool
wing: mottled woodcock quill section tied flat and V-notched at the tip
hackle: light ginger

If woodcock is unavailable for the wing, hen pheasant may be substituted. Incidentally, the only way you may be able to obtain woodcock is if you hunt, or have a friend who does and is willing to save you the feathers.

Nat's Caddis is also tied in a darker shade in which the body is made of dun grey wool and the hackle dark ginger.

SAM MCALEES fishes the Rockies in northern New Mexico with his friend Bob Pelzl. Both reside in Albuquerque. The Troth Elk Hair Caddis and the Caddis Larva were tied by Bob, while Sam submitted the Caddis Pupa, which is an imitation of *Rhyacophila grandis.* The three patterns have been used effectively over the past five years in the small mountain streams of both New Mexico and Colorado.

troth's elk hair caddis *(tied in size 16 or larger)*

hook:	Mustad 94833
thread:	Herb Howard tan
body:	blend of one part bleached raccoon and one part hare's ear fur
rib:	fine gold wire
hackle:	dark ginger palmer through body
wing:	elk hair, tan
head:	butt ends of elk hair wing clipped in front of tie down area
note:	Pattern is also tied in olive and dark brown body colors

caddis larva *(tied in sizes 14 to 18)*

hook:	Herter's No. 707
thread:	Nymo, brown
abdomen:	bleached raccoon
thorax:	any darker fur; tan or brown
covert:	brown polypropylene yarn
rib:	buttonhole twist, shade #3010
hackle:	furnace or dark ginger

note: This pattern is also tied in grey and green. It is usually weighted with fine lead wire.

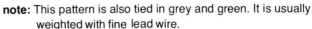

caddis pupa *(tied in size 14)*

hook:	Mustad 9671
thread:	Herb Howard tan
body:	blend of one part olive seal and one part olive Poly II over an underbody of kelly green silk floss.
legs:	dark partridge
wing:	double blue dun hackle tips
antennae:	single strands of wood duck
head:	dark brown emu (mini ostrich can be substituted)

DAVE McNEESE, of Vaneta, Oregon, who so generously contributed help to complete Chapter 6, mailed the authors the pattern listings of the caddis imitations he uses throughout the season; most of them do not have any names. That is unimportant; size and color are the primary factors. Here are Dave's pupa and adult imitations.

brown/grey caddis pupa *(tied in sizes 10 to 18)*

body:	brown fur
thorax:	dark brown fur
wing case:	dark slate duck quill
legs:	dark brown partridge hackle

light brown/dark grey caddis pupa
(tied in sizes 12 to 18)

body:	light brown fur
thorax:	brown (darker than body)
wing case:	dark slate duck quill
legs:	creamish mottled brown

dark cream/grey caddis pupa *(tied in sizes 12 to 18)*

body:	dark cream fur
thorax:	buff colored fur
wing case:	grey duck quill
legs:	creamish mottled brown

olive/grey caddis pupa *(tied in sizes 12 to 18)*

body: medium olive green fur

thorax: dark olive brown fur

wing case: grey duck quill

legs: buff/grey mottled

brownish grey/grey caddis pupa
(tied in sizes 14 to 18)

body: light brown mixed with grey dubbing fur

thorax: brown fur

wing case: grey duck quill

legs: light brown mottled

orange brown/dark grey caddis pupa
(tied in sizes 14 to 18)

body: orange/light brown dubbing mixed

thorax: dark brown fur

wing case: dark grey duck quill

legs: grey partridge dyed pale orange

dark brown/black caddis pupa
(tied in sizes 14 to 18)

body: very dark brown, almost black dubbing

thorax: black fur

wing case: black

legs: dark brown mottled grouse

dark brown/black caddis *(tied in sizes 12 to 18)*

body: dark brown fur

wing: very dark brown dun tied half again as long as body; stiff hackles or deer hair are used

hackle: dark rusty dun set behind wings; iron dun set forward of wings

brown/rusty dun caddis *(tied in sizes 12 to 18)*

body: brown fur

wing: rusty dun hackle fibers or equivalent in deer hair

hackle: dark bronze dun behind wings; medium rusty dun in front

mc kenzie green caddis *(size 12)*

body: greenish-pale blue dubbing
wing: medium dun grey deer hair
hackle: grizzly mixed with dyed green hackle

orange caddis *(tied in sizes 6, 8)*

body: pale orange fur
wing: natural dun bucktail
hackle: grizzly mixed with dyed orange hackle

greyish olive/dun caddis *(tied in sizes 12 to 18)*

body: greyish olive fur
wing: dark bronze hackle fibers or deer hair equivalent
hackle: bronze dun, before and after wings

olive/grizzly cream caddis *(tied in sizes 8 to 18)*

body: medium olive fur
wing: grizzly and cream hackle fibers mixed or equivalent effect with light-colored deer hair
hackle: grizzly and cream mixed, both behind and in front of wing

buff/brown caddis *(tied in sizes 12 to 16)*

body: pale buff fur
wing: light brown dun hackle or deer hair equivalent
hackle: dark cream and medium blue dun mixed; tied both in back and front of wing

Whenever JACK MICKIEVICZ gets skunked he dredges the stream and combs the air for insects trout have apparently been feeding on. Regardless of how tired he is, or how long the day has been, he will sit down at the vise the very same evening, his specimens before him, and try to determine just what it was the trout were taking. He always comes up with something, and it will be presented to the same trout the very next day. If they still refuse it Jack begins the process all over again. Now that's innovative fly tying.

Jack's partner in crime is DICK ESTLER, and they both live in Phoenixville, Pennsylvania. In addition to the patterns and their descriptions, Jack also sent along some dozen odd sheets of pertinent information that has been a useful resource throughout this book.

The very first pattern listed is one which Jack has been using for years. He describes the tying procedure: "It's so simple, it's ridiculous." It is simple, but deadly. It is called the Honey Bug.

honey bug (caddis larvae)
(tied in all larval sizes)

hook: Mustad 3906, 3906B, 37160

body: cotton chenille (tied in appropriate colors)

There you have it. On occasion Jack will use a marking pen to produce a darker back (two toned effect), for the larval imitation. The only other observation I have to add is that most caddis larvae have black heads. For that reason I would not hesitate to add a wrapping of peacock or ostrich herl just in back of the eye of the hook. Jack wraps the cotton chenille (which is softer than rayon chenille) to the eye and hides the thread with the overlapping fibers.

quill caddis pupa *(tied in size 12)*

hook:	Mustad 9479 or equivalent 5X short, down eye
thread:	olive
tail:	two peacock fibers (very short)
underbody:	green wool
body:	translucent flue from short side of black and white primary Turkey flight quill
hackle:	hen pheasant shoulder feather (brown and tan), tie sparse (two turns), extending to tail
antennae:	two wood duck flank fibers tied long
head:	greyish brown mole

Jack contends that this pattern is effective on nearly any emerging caddis fly hatch.

dry caddis emerger *(tied in sizes 12 to 24)*

hook: mustad 94840 or equivalent

thread: olive or orange

body: grey squirrel body with some guard hairs mixed in

legs: baby seal or substitute to color body as desired

wing: grey squirrel guard hairs

hackle: none

This pattern is very fuzzy and buggy in appearance. Of course all of Jack's patterns should be related to color and size of the species emerging.

muddy water rock worm *(tied in sizes 10 and 12)*

hook: Mustad 9672 bent, or 37160

thread: olive

tail: 3 peacock sword fibers (short)

body: fluorescent kelly green baby seal or substitute

rib: dark olive polyester thread

head: peacock herl

Jack writes: "This pattern is used only when the water is fairly clouded or actually muddy; it won't work in clear water."

olive caddis pupa *(tied in size 12)*

hook: Mustad 9479

thread: olive

body: light olive seal, green rabbit, grey squirrel, and amber seal mixed in equal parts

legs: hen pheasant neck

wing: mallard wing quill (trimmed to half body length and sprayed with vinyl cement)

eyes: 8-pound test monofilament burned at each end to form stubs, or "eyes"

This is the only pattern I've come across in which a tier has put eyes on his imitation. The eyes do stand out. Does this addition increase the number of strikes? There is only one way to find out.

little green caddis pupa *(tied in size 18)*

hook:	Mustad 7948A
thread:	olive
body:	bright insect-green rabbit
rib:	olive polyester thread
thorax:	bright insect-green rabbit
wing:	black duck wing quill (half body length)
legs:	dark woodcock or quail

cinnamon caddis pupa *(tied in size 16)*

hook:	Mustad 7948A
thread:	orange
body:	grey Australian opossum, dyed yellow
rib:	fine gold thread
thorax:	grey Australian opossum, dyed yellow
wing:	mottled secondary male ringneck pheasant tail
legs:	woodcock or quail rump

brown caddis *(tied in size 12)*

hook:	Mustad 9479
thread:	brown
body:	red fox, grey squirrel, and brown/olive seal fur mixed in equal parts
rib:	brown polyester thread
wing:	mottled secondary male ringneck pheasant tail
legs:	hen pheasant hackle

In the foregoing list of pupa imitations notice that Jack was meticulous to the point of varying size and color very specifically. In other words, he did not describe a brown caddis for style and then suggest other sizes and colors in order to imitate different species. Also, the style of hooks used varied from pattern to pattern.

Jack is a proponent of the blended fur formula. (I agree with him.) He does not use a solid color if he can arrive at the shade by using any number of blends to attain the desired effect. Colors which have been attained by dyeing are of one single color. Colors which are attained by blending appear to be of one single color, *but have many other colors inherent in them,* thus making for an imitation which is closer to the natural (insects are never one solid color).

In the adult (dry fly) category Jack uses two Wulff type of patterns for his fishing. They read as follows.

hare's ear wulff *(tied in sizes 8 to 14)*

hook: Mustad 94840 or 94833
thread: olive
tail: English hare jowl
body: remnants from tying in wing and tail
wing: guard hairs from English hare's mask
hackle: grizzly, cree and medium dun mixed

caddis wulff *(tied in sizes 8 to 14)*

hook: Mustad 94840, 94833
thread: dark brown
wing: grey squirrel tail
tail: grey squirrel tail
body: half rusty brown Australian opossum, one-quarter olive green rabbit, 25% yellow rabbit
hackle: two cree and one grizzly mixed

If you are going to tie any of Jack's flies I suggest you buy a blender.

The use of Wulff types of flies for caddis is not new. They are, in fact, very effective at times, especially in turbulent waters. They ride high and can be twitched effectively.

Some of the more currently popular caddis patterns tied today employ the use of mink tail guard hairs in their construction. The use of mink fibers began as a substitution for the hackle wing of the Leonard Wright Fluttering Caddis since it was sometimes difficult for the average tier to obtain the long, stiff hackle fibers called for in the original. Mink tail guard hairs are actually stiffer and longer than hackle fibers, and they are available in a variety of natural, bleached, and dyed colors.

Here are some of the patterns known as the Mink Wing Caddis.

brown mink wing caddis *(tied in sizes 12 to 18)*

hook: Mustad 94840
body: brown
wing: brown mink tail guard hairs
hackle: brown

grey mink wing caddis *(tied in sizes 14 to 20)*

hook: Mustad 94840
body: grey muskrat or mink dubbing
wing: grey mink tail guard hairs
hackle: medium blue dun

olive mink wing caddis *(tied in sizes 14 to 20)*

hook: Mustad 94840
body: olive dubbing
wing: brownish grey mink tail guard hairs
hackle: blue dun

olive/tan mink wing caddis *(tied in sizes 16 to 22)*

hook: Mustad 94840
body: olive/tan dubbing
wing: tan mink guard hairs (or bleached to dark ginger)
hackle: dark ginger

When I asked JIM GILLIS, a very knowledgeable angler from South Bend, Indiana, for a favorite caddis pattern, he replied, "Frankly, I prefer the mink wing version and fish them as a fluttering caddis. They work for me wherever I go."

"Hey Tom, what's the name of that pattern you're always raving about?" I could hear ROGER MISTOVE's voice through the phone as he called to one of his fly-tying friends. The next voice to come through the receiver was that of TOM GRIFFITH. Tom and Roger are both from North Plainfield, New Jersey. It seems that Tom has had much success using a caddis larva of his own design, which he graciously outlined for me over the phone. Here is the pattern.

griffith caddis larva *(tied in sizes 10 to 16)*

hook: English bait hook or equivalent
body: white polar bear dubbing (or substitute white Sealex)
rib: white cotton thread
thorax: natural white seal fur mixed with brown beaver fur
legs: partridge hackle

When someone raves about how many fish he takes with a pattern, you try it. I will.

During one of my visits to Michigan, TOM NAUMES of Bellaire, invited me to fish the Manistee River. It was during the hatch of the giant Michigan mayfly, *Hexagenia limbata,* late in June. Three other anglers, in addition to Tom and myself, were stationed at various locations on the stream awaiting the emergence which takes place after dark. All of us were skunked—except Tom, who took and released six nice fish. He gave me an actual demonstration of how it was done; unfortunately, it was too dark for me to see what he was doing!

I asked Tom if he would send a listing of some of his favorite caddis patterns. They are listed below.

brown caddis pupa *(tied in sizes 10 to 16)*

hook:	Mustad 3906B
body:	brown seal fur or Sealex
rib:	silver wire
wing:	slate duck quill tied short on sides of shank
hackle:	grey partridge

black caddis—dry *(tied in sizes 14 to 18)*

hook:	Mustad 94840
body:	black Fly Rite polypropylene
wing:	black mink guard hairs
hackle:	black

grey larva *(tied in sizes 10 to 16)*

hook:	Mustad 3906B
body:	natural grey rabbit
legs:	black deer hair
head:	black wool (or peacock herl)

Note: This pattern is tied weighted.

GUS NEVROS builds bamboo fly rods in the tradition of Jim Payne and Everett Garrison. He does not build many, but they are works of art—beautiful, delicate, and balanced. Each rod is the result of an inborn love for perfection. In his home in Manhasset, New York, Gus pursues many other angling arts, among them, tying flies.

One day while we both fished Esopus Creek, a stream flowing easterly along the northern sector of the Catskill Mountains, Gus handed me a caddis imitation. It appeared to be similar to many others I had fished, except for the body, which was made of a lacquered red quill. The same type of red quill that is used in the Red Quill pattern described by Art Flick. Therein was all the difference.

We both fished this fly during the grannom hatch *(Brachycentrus numerosus)* though there are many patterns which more closely resemble this insect. Perhaps it is the delicate, almost ephemeral makeup of this pattern (it does not hold up too long) that causes it to be so effective. Whatever the reason, the pattern does take fish extremely well. I've named the fly in his honor.

gus's grannom *(tied in sizes 12 to 16)*

hook:	Mustad 94840
body:	lacquered quill stem from Rhode Island red rooster
wing:	mallard primary, cupped and lacquered
hackle:	brown or pale rusty dun

ED and RICK OLIVER of Clinton, New Jersey, have been good friends for a long time. I never fail to receive a hearty welcome during my visits there. When I asked for one of their favorite caddis patterns it arrived the next day.

In Rick's words, "This pattern has proven to be very successful in early and mid-season on most New Jersey streams. It is also a killer on the Big Bushkill and Broadheads in the Poconos. It represents the common cased caddis out of its case. Duck quill wings can be added to imitate pupae. Fly should be fished on or near the bottom. The fly has become very popular because it is so simple to tie, but mostly because it produces fish."

This is a larval imitation called the Sand Caddis.

sand caddis *(tied in sizes 10 to 14)*

hook:	Mustad 9671 or 37140 (weighted)
body:	bleached tan opossum fur
rib:	fine gold wire
thorax:	black fur picked from hare's ear
legs:	dark partridge hackle
head:	black

DR. FRED OSWALT, also of Michigan, resides in Battle Creek. He collects antique clocks and watches but never seems to know what time it is, especially when he is lost astream or creating new flies in his den. He has come up with a pattern designed to imitate the emerging caddis which occasionally gets stuck in its shuck. After much thought he discarded the name "stuck shuck" caddis in favor of the pattern named and described below.

crippled caddis *(tied in sizes 12 to 18)*

hook:	Mustad 94840
tail:	trailing sheath of Swiss straw (two thirds body length)
body:	polypropylene
wing:	Swiss straw slanted to abdomen and trimmed to shape
hackle:	rooster hackle for legs

This is a general caddis pattern. Color used should be that of natural being imitated. Fred advises drag free presentation or "the fly won't work".

I don't know of anyone who has ever met DR. WILLIAM R. (Bill) PRIEST, of Saginaw, Michigan, who has not said in essence, "What a great guy!" or, "He's something else!" Bill has a fine reputation among anglers, and deservedly so. He is a tier, teacher, student, angler and worker—especially a worker. Many of us talk about doing something for clean waters and better fishing, and perhaps one day we will—but Bill is one of those who does.

When I asked Bill for a few caddis patterns useful in Michigan streams, I received not only a half dozen of his own field-tested favorites but an equal number of pages as well describing them. So in reporting them I shall use some of his words, as he expressed them.

The first one, the Little Black Caddis "produces some of the best fishing on our streams after the Hendricksons have tailed off. The fish seem to prefer the fly as it emerges from the pupal shuck and the pattern should be fished in the surface film. *Do not use a floatant with these patterns*" (the first three flies listed).

little black caddis *or* chimarra caddis

(tied in sizes 16 and 18)

hook:	Mustad 94840 or equivalent
body:	black horsehair, or zebra mane
wing:	mouse deer hair (substitute: hair from mask of deer; doe is excellent.)
hackle:	black

Bill writes, "Lately some of the guys have been using the dubbed poly body with good success."

chimarra caddis #2

body:	black poly dubbing
wing:	doe face hair
hackle:	black

chimarra caddis #3

body:	black poly dubbing
wing	trimmed wing made from sheet of poly II material light grey in color
hackle:	black

On the Little Black Caddis #3 the Poly II wing is sprayed with Krylon before it is cut from the sheet. This makes it easier to form with the scissors. Bill goes on to note; "Once the wing has been cut, crease it down the middle and tie it over the hackle as you would a jungle cock nail when tying a Jassid. It is important to reinforce the crease with vinyl cement or else the wing will hinge away from the body and cause the tippet to twist."

At this point in his letter, Bill gives us a tip on fishing the first three patterns. "Do not be in a hurry to pick up the fly as it begins to drag; as a matter of fact, I frequently allow the fly to hang in the current and slowly sweep the rod tip back and forth before picking it up for the next cast. *Keep your hands off the reel!* Point the rod at the fly while doing this unless you want to lose your fly. Trout really attack these emerging flies and I've popped many tippets when I've been careless. *Try and let the fish strike with only the reel drag as resistance.*

"Pattern number 3 is a 'skittering caddis' tied with a poly wing. It works well when the adult flies are fluttering about. . . ."

skittering caddis *(tied in sizes 14 to 18)*

hook: Mustad 94840 or equivalent

body: grey fur or poly dubbing

wing: white poly yarn tied spent

hackle: brown and grizzly mixed

Bill seems to feel that the adults that are seen fluttering about are laying eggs, hence the pattern is tied with a spent wing, which is similar to the Delta Wing described in Chapter 4. Adult caddis do flutter during ovipositing, but also during emergence.

The fifth pattern listed by Bill is a very old one and has been used on the North Branch of the Au Sable river for many seasons. It is called the Harris Special.

harris special *(tied in sizes 12 to 18)*

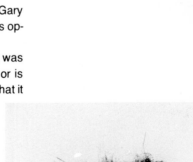

body: yellow tying thread, flat (monocord) (Body color may be olive, grey, or ginger)

wing: guard hairs from grey squirrel

hackle: brown and grizzly

thread: olive

This pattern is very similar to the one sent in from Wisconsin by Gary Borger, called the Eddy Fly. The Eddy Fly uses woodchuck guard hairs as opposed to the squirrel tail guard hairs for the wing.

The last of the patterns Bill sent is one he calls the Cased Caddis. It was created trying to imitate the cased larva; however, Bill suspects that it is or is suggestive of something else. Though he "hesitates to include it" he adds that it has been a good producer during caddis activity.

cased caddis *(tied in sizes 8 to 14)*

hook: Mustad 3906B

body: dubbed fox squirrel body with guard hairs left in. (Excess guard hairs are trimmed with scissors)

head: black fur

Overall effect is bristly and shaggy.

Bill concludes with the comment that flies number one and four, the Little Black Caddis, are the patterns he developed but "like any honest fly tier, I've borrowed ideas from many friends and fellow tiers."

POLLY ROSBOROUGH of Chiloquin, Oregon, and author of *Tying and Fishing the Fuzzy Nymphs* tied two flies on the spot during a Federation of Fly Fishermen conclave in Sun Valley, Idaho. "This is what you want to use when those big caddis come off," were his words.

light ginger caddis *(tied in sizes 2 to 8)*

hook:	Mustad 94840 or equivalent
body:	tan orange fur or wool
rib:	palmered golden ginger hackle
wing:	buff ginger deer hair, or equivalent

dark caddis *(tied in sizes 6 and 8)*

hook:	Mustad 94840 or equivalent
body:	dirty orange wool or fur
rib:	very dark brown hackle
wing:	dark brown deer hair or equivalent
hackle:	very dark brown

GENE SNOW, of Salt Lake City, Utah, sent what at first appeared to be a conventional Elk Hair Caddis. Upon closer inspection of his drawing and description, I realized that he was using flared elk hair which is trimmed (almost as you would the head of a Muddler Minnow), so that the fly rides close to the water, yet is durable and representative. Gene claims this pattern is destined to become one of the standard caddis versions used in the Rockies; here is its description.

elk hair wing caddis *(tied in sizes 8 to 16)*

hook:	Mustad 94840
body:	peacock herl, or substitute for other colors
wing:	elk hair (tied back fluttering style)
hackle:	elk hair spun and flared, trimmed to shank on bottom of hook; elk hair fibers left at sides of shank to represent legs

LARRY STRAWN is a professional tier from Salem, Oregon who enjoys a little innovating. The pattern listed below is tied parachute-style. It works.

parachute caddis buck (gold) *(tied in sizes 8 to 16)*

hook:	Mustad 94840 or 94833
body:	gold poly yarn
wing:	elk or deer body hair
hackle:	brown, about one size larger than hook tied as a parachute
head:	lacquered black

An optional version is to rib the body with a palmered orange hackle. Yellow thread is used throughout the tying operation.

In listing the foregoing patterns sent in to the authors I have tried to note all the information and any brief histories that were available. In some cases flies other than those originated by the tiers themselves were recommended as favorites; if their creator was known proper credit was given.

For the most part, these files can be tied by their description alone. However, if any problem should arise in their construction please don't hesitate to write the authors for further information.

All mail will be answered, though there are times the answers are a bit delayed due to an overload of field testing to be done on the stream . . . especially during good weather and rising trout.

we are pledged

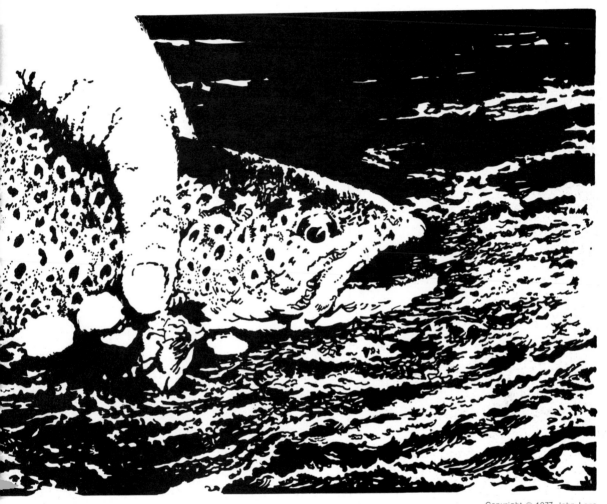

to trout release

appendices

sources of supply

The following listing of retail mail order houses dealing in flytying materials and fly fishing equipment can be relied on in the event that you cannot obtain the merchandise you require from your local fly shop or sporting goods dealer. It is always better to be able to see what you are buying before you buy it, but mail order houses can only survive in business by their reputation. Those folks will take good care of you. If you have a complaint in dealing with any of them of any nature, let them know about it, so that they can rectify any mistake. All of those listed are reputable and will do their best to please you.

Dan Bailey's Fly Shop
209 West Park St.
Livingston, MT 59047

Beckie's Sporting Goods
1336 Orange St.
Berwick, PA 18603

Bodmer's Fly Shop
2400 Naegele Rd.
Colorado Springs, CO 80904

Buszek's
805 West Tulare Ave.
Visalia, CA 93277

Cascade Tackle
2425 Diamond Lake Blvd.
Roseburg, OR 97470

Dale Clemens Custom Tackle
Rte. 3, Box 415F
Allentown, PA 18104

Creative Sports Enterprises
2333 Boulevard Circle
Walnut Creek, CA 94595

E.B. and H.A. Darbee
Livingston Manor, NY 12758

Fishin' Fool Fly Shoppe
760 Main Avenue N.
Twin Falls, ID 83301

The Fly Fisher
315 Columbine St.
Denver, CO 80206

Fly Fisherman's Bookcase
    and Tackle Service
3890 Stewart Rd.
Eugene, OR 97402

Fly Fisherman's Headquarters
169 Route 46 (Mine Hill)
Dover, NJ 07801

Fly Tyer's Supply Shop
Box 153
Downingtown, PA 19335

Hackle & Tackle
553 North Salina St.
Syracuse, NY 13208

E. Hille
815 Railway St.
Williamsport, PA 17701

Hook & Hackle
Box 1003
Plattsburgh, NY 12901

Bob Jacklin's Fly Shop
West Yellowstone, MT 59758

Jack's Tackle
301 Bridge St.
Phoenixville, PA 19460

H. L. Leonard
25 Cottage St.
Midland Park, NJ 07432

Bud Lilly's Trout Shop
West Yellowstone, MT 59758

The Orvis Company
Manchester, VT 05254

Patrick's Fly Shop
2239 Eastlake E.
Seattle, WA 98102

Rangeley Region Sports Shop
Rangeley, ME 04970

Reed Tackle
Box 390
Caldwell, NJ 07006

Raymond C. Rumpf & Son
Box 176
Ferndale, PA 18921

S&M Fly Tying
95 Union St.
Bristol, CT 06010

Shoff's Tackle Supply
Box 1227
Kent, WA 98031

Streamborn Flies
13055 S.W. Pacific Hwy.
Tigard, OR 97223

Tack-L-Tyers
939 Chicago Ave.
Evanston, IL 60202

Thomas & Thomas
22 3rd St.
Turners Falls, MA 01376

Twin Rivers Tackle Shop
1206 N. River Dr.
Sunbury, PA 17801

Universal Imports
Box 1581
Ann Arbor, MI 48106

E. Veniard, Ltd.
138 Northwood Road
Thornton Heath CR4 8YG
England

Yellow Breeches Fly Shop
Box 200
Boiling Springs, PA 17007

selected bibliography

The following publications will prove useful to the angler, flytyer and amateur entomologist, as they have to the authors.

Bay, Kenneth E. *How to Tie Fresh Water Flies.* New York: Winchester Press, 1974

Betten, Cornelius. "The Caddis Flies, or Trichoptera of New York State." New York: New York State Bulletin, 1934

Boyle, Robert H., and Whitlock, Dave. *The Fly Tyer's Almanac.* New York: Crown Publishers, Inc., 1975

Brooks, Charles. *Nymph Fishing for Larger Trout.* New York: Crown Publishers, Inc., 1976

Caucci, Al and Nastasi, Bob. *Hatches.* New York: Comparahatch, Ltd., 1975

Flick, Art. *Art Flick's Master Fly Tying Guide.* New York: Crown Publishers, Inc., 1972

Jorgensen, Poul. *Dressing Flies for Fresh and Salt Water.* New York: Freshet Press, 1973

———. *Modern Fly Dressings for the Practical Angler.* New York: Winchester Press, 1976

Kreh, Lefty. *Fly Casting with Lefty Kreh.* Philadelphia and New York: J. B. Lippincott Company, 1974

LaFontaine, Gary. *The Challenge of the Trout.* Missoula: The Mountain Press, 1976

Leisenring, James E., and Hidy, Vernon S. *The Art of Tying the Wet Fly and Fishing the Flymph.* New York: Crown Publishers, Inc., 1973

Leiser, Eric. *Fly Tying Materials.* New York: Crown Publishers, Inc., 1973

Needham, Paul R. *Trout Streams.* New York: Winchester Press, 1969

Ross, Herbert H. 1944. "The Caddis Flies, or Trichoptera of Illinois." Urbana: *Bulletin of Illinois Natural History Survey:* 23.

———. *Evolution and Classification of Mountain Caddis Flies.* Urbana: University of Illinois Press, 1956

Schwiebert, Ernest. *Nymphs.* New York: Winchester Press, 1972

Swisher, Doug, and Richards, Carl. *Fly Fishing Strategy.* New York: Crown Publishers, Inc., 1975

———. *Selective Trout.* New York: Crown Publishers, Inc., 1971

Wright, Leonard. *Fishing the Dry Fly as a Living Insect.* New York: E. P. Dutton, 1972

index

222